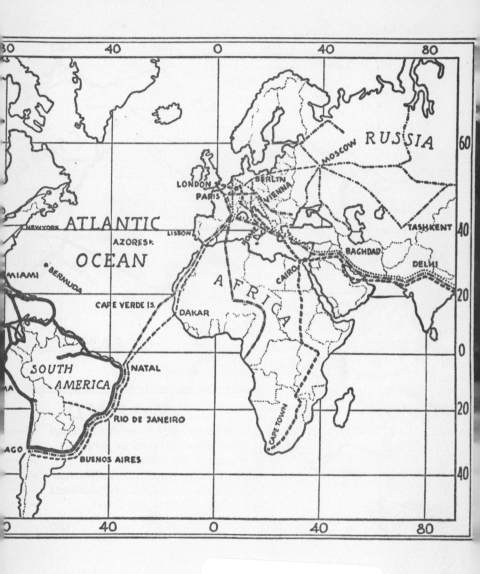

SKYWAY TO ASIA

SKYWAY TO ASIA

BY

WILLIAM STEPHEN GROOCH

LONGMANS, GREEN AND CO.

NEW YORK · TORONTO

GROOCH
SKYWAY TO ASIA

First Edition September 1936
Reprinted September 1936
October 1936, December 1936
February 1937, November 1939

To

IRENE GROOCH

and

FRANK WEAD

FOREWORD

THIS is a story of the first North Haven Expedition, dispatched to build commerical air bases across the Pacific Ocean — the stepping stones for the flying Clipper Ships on their airway to the Orient.

It is a story of personal experiences, my own — not an official record of the Expedition. Nor is it a story of the Clippers themselves, of their design and building and their sky voyages, nor of the moves that led to the conception of this singular American enterprise.

This book will come as a great surprise to my associates who were not consulted in its preparation. The last thing they and the Company would expect from me would be a book!

This book salutes them all in my name.

Alameda Airport
June 1936

LIST OF ILLUSTRATIONS
to be found at end of volume

China Clipper

North Haven loading at San Francisco

Discharging cargo at Midway Islands

"Bread Line" during first days at Midway Islands

Temporary radio shack at Midway

Baby Gooney

Constructing Mess Hall at Midway Islands

Love birds, Midway

Baby love bird on nest

Parent Gooneys gossiping

Young Gooneys on the beach on Eastern Island

Calibrating the radio compass

Five-ton generators going ashore at Wake Island

Gooney Lodge initiation ceremonies

Preparing for initiations into the Ancient and Royal Order of Gooneys, on the *North Haven*

Constructing a dock at the boat landing Wilkes Island

Clearing a right of way for the railroad on Wilkes Island

The railroad ending at the Lagoon

The railroad in action

Clearing Wilkes Channel to bring lighter through

Pushing the lighter through Wilkes Channel

SKYWAY TO ASIA

Tractor in Wilkes Channel hauling the lighter off a shallow spot

The light launch is put overboard into the lagoon

Mullahey with his fishing sling and arrow

Wingless rail attacking a hermit crab

Tractor arriving at Peale Island

Mess hall and cold storage on Peale Island

The main highway — Guam

Fast transportation — Guam

Looking down on the station on Peale Island from the top of the wind mill

Company Street at Peale Island

SKYWAY TO ASIA

SKYWAY TO ASIA

CHAPTER ONE

FLYING down to Rio! The picturesque old city of Bahia, which we had just left, faded from sight astern as the big twin-engined *Commodore* droned southward toward Rio de Janeiro, nine hundred miles away. It was easy flying, perfect weather and the course followed the coastline. My passengers were settling down for a morning nap. I reflected that the job was easy now.

But four years before, in 1929, I had flown the first *Commodore* from Buffalo, New York, to Buenos Aires, down the same coast. It was all strange country to me then. I couldn't speak a word of Portuguese, and there were virtually no flying facilities.

I had just resigned from the Navy to accept a position as operations manager for a new air line, The New York, Rio and Buenos Aires Line, known as Nyrba to its intimates.

The months that followed led me to believe that pioneering an air line is the toughest racket on earth. Many times I wished myself back in the Navy where for twelve years I had had a good time and few worries. The Navy taught me to fly during the World War.

Nyrba made fine progress for a while. Passenger, mail and express loads slowly increased. But then the market crash of 1929 began to take effect and Nyrba almost died of starvation. Pan American Airways, which had also been reaching south from Miami, down the east and west coasts of South America, came to our rescue. They took Nyrba over lock, stock and barrel.

I joined Pan American's staff. The going was much better then, but after four years I had become a bit bored with checking off the lighthouses between the Amazon River and Buenos Aires, and had requested a change of duty. As we droned southward from Bahia I wondered what my new assignment would be.

Twenty miles out of Rio the steward notified all passengers that we would soon be in. Each promptly glued his face to a window.

Flying at five thousand feet we cut across a range of mountains that jutted out into the sea. The great silver bowl of Rio harbor unfolded below us. The twelve passengers gasped at the beauty of the scene.

Shoulder to shoulder the giant hills stand guard around the shores of the harbor, leaving only a narrow gate for the entrance. Far to the westward a row of organ pipes flare against the sky-line, the tallest being "Dedo de Deus" (Finger of God). To the south, "Corcovado" (The Hunchback) Mountain, with a hundred-foot-high statue of Christ, looks down from three thousand feet on the gayly colored city. Rio is built on an uneven shelf at the foot of the mountain. Its suburbs wander over numerous small hills.

We slid down past Sugarloaf, which flanks the entrance to the harbor, and circled over a small island just across the bay from the city. Here were the small white buildings of the Pan American seaplane station.

The checkered flag was flying, which meant everything was O. K. I landed and taxied up to the float.

As I stepped ashore the airport manager shook hands and told me to phone the operations manager. As soon as I had checked in I called Hump Toomey.

"Bill," said Hump, "stop in a minute on your way home. I've got some news for you."

At Hump's office he handed me a radio message from Pan American's New York headquarters. The message read: "Have recommended Grooch for duty as operations manager in China stop If he accepts must take next plane for New York."

For four years I had flown over the jungles, rivers, coasts and mountains of most of the countries in South America. I was due for a change of scenery. China sounded good to me. I caught that next plane for New York.

A week later I sat across the desk from Pan American's chief engineer, André Priester, in the New York office. The Chief doesn't talk much but what he had to say gave me the surprise of my life.

"Grooch," he said, "we are going to install an airline in China from Shanghai to Canton. You are being sent as operations manager. You will sail from Los Angeles on the Maersk Line in about a week, so you haven't much time to waste."

"I'll be ready," I answered. "But what on earth are you going to do with an air line in China?"

"Do you know much about the Orient?" he asked.

"Practically nothing," I admitted.

"America does as much business with the Orient as with all of South America," the Chief stated. "China needs air transportation badly, and so do the Philippines. When things are running smoothly on the line from Shanghai to Canton, the next step will be to extend it from China to the Philippines."

"Fair enough," I said. "But all the rest of our lines are hooked up with the States. We've got American mail and American passengers to depend upon for business. How are we going to make a go of it with the Pacific Ocean between us?"

There was a brief silence while the Chief appeared lost in thought. Then with a slow smile he said, "Maybe it's possible to build a bridge."

I stared at him in amazement. "You mean to fly the Pacific? It's eight thousand miles to Manila!"

"Well," he answered, "it's eight thousand miles to Buenos Aires, and we fly it every week."

"Sure we do," I replied, "but we've got a coastline under us or islands where we can stop and refuel."

"There are islands in the Pacific."

"Oh! By way of the Aleutians?"

The Chief shook his head. "Honolulu," he said, shortly.

Honolulu! Twenty-four hundred miles from California. In flying, pilots often speak of a long flight as a "big hop." This was a big hop, all right; the longest over-water jump any air line ever dreamed of.

"Has some one invented an engine that will run without gas?" I asked.

"No," he said. "But we have some plans coming along that will surprise you."

I asked one more question while he was in the mood to talk. "What about air bases west of Honolulu?"

"Midway Islands, Wake Island, Guam and Manila," he answered.

I made a mental note to take a refresher course in geography right away.

The next day he took me over to the Sikorsky plant at Bridgeport, Connecticut, where the new, big Sikorsky Clipper S42 was under construction. Four big motors and a wing spread of one hundred and fourteen feet. It wasn't very far along then but there was enough of it in sight to indicate that here was something bigger and different in flying boats.

I tried to find out more about the transpacific run but the Chief was noncommittal. He made it clear that the plans were being carefully worked out and that for the present my job was in China, which would give us a foothold at the eastern end of the line.

A week later I flew to Los Angeles and boarded the

steamer *Gertrude Maersk* for Shanghai. Planes, pilots, mechanics and radio men for the China-coast run were already on board. Our first stop was Yokohama.

In Japan the officials were very curious about our mission, and asked innumerable questions.

We arrived in Shanghai on June 26, 1933. Pan American's Far Eastern Manager, H. M. Bixby, met us at the dock. He, by the way, was the principal sponsor of Lindbergh's flight to Paris.

The mechanics made a fast job of setting up our first plane. On July 3 I started on a survey flight down the coast of China. The first stop was Wenchow, a queer old town two hundred and sixty miles below Shanghai. The native boatmen swarmed about us as we anchored in the river, and we had great difficulty in preventing damage to the plane.

Mr. Bixby and I inspected the town. It was lucky we had an interpreter along, as we found no one who could speak English except a Chinese school-teacher. We decided to make him our local agent temporarily.

There was not a street in the town wide enough to accommodate an automobile. Quite often rickshaws had to draw into a doorway to let other rickshaws pass.

After a Chinese luncheon, we flew on to Foochow. There we were met by the young American Vice Consul, Tom Reynolds, who put us up for the night.

Foochow lies thirty miles up the Min River from the coast, and is quite an important city. It is entirely surrounded by mountains, and the valley is green and fertile.

Flying low down the coast from Foochow we could see the farmers ploughing their rice fields with water buffalo. Man and beast plodded along in mud and water knee-deep, with a forked stick for a plough.

Hong Kong, our next stop, is built on the side of a hill. It is a lovely sight at night when the lights are turned on. Next to Shanghai it is the most important city of China.

From Hong Kong we went on to Canton, which lies in flat country ninety miles up the Pearl River. Near the city the river was choked with thousands of junks and sampans, and there was no clear space for the plane to land. We had to go back eight miles down-stream before we dared "sit down."

We paid the customary official calls, and informed the Chinese post office that we were ready to fly the mail from there to Shanghai. Much to our surprise the postmaster stated he had received orders to hold the matter in abeyance until further notice. It appeared that some Chinese officials had failed to sign our franchise in the proper place. Red tape is unusually baffling in China, and there is no end of it.

We flew back to Shanghai and found that it would take several weeks to straighten out the tangle. Pan American decided to employ this enforced delay to best advantage. They directed me to equip one of our planes for a flight across the China Sea to Manila.

This flight was attended by much local publicity, largely because some amateur pilot had attempted it a few months earlier, and failed to arrive.

I made the flight without difficulty, and landed on Manila Bay in front of the Army and Navy Club. Our reception was most cordial, and the Press treated us handsomely.

Everyone was enthusiastic at the news that Pan American was planning an air line from Hong Kong to Manila. But they were politely incredulous when we promised them a transpacific air line within two years. It was difficult for me to make the story convincing because no matter how I worded it it did sound like a fantastic dream.

A visiting Englishman cornered me one night at a club bar and stated bluntly that he thought this talk of a transpacific air line was pure nonsense.

A bit nettled, I replied: "Would you care to make a friendly bet that we won't fly a load of mail across the Pacific two years from now?"

"My dear man," he snorted, "I have traveled by air all over Europe and over the Dutch line to Java. I concede that your company can probably manage to fly from Hong Kong to Manila. But an air crossing of the Pacific Ocean —the idea is absurd."

"I suppose that's an average opinion," I answered. "Aviation is not your profession, and you've never made a study of this particular problem. Colonel Lindbergh has. He is technical adviser to Pan American, and has approved our plans for flying the Pacific."

The Englishman just stared at me.

During our two weeks' visit in Manila, I made an air survey of the Philippine Islands as far south as Zamboango.

On our return flight to China we passed over the Dollar Liner *President Hoover,* and dropped a bundle of Sunday newspapers on her deck. Captain Anderson of the *Hoover* thanked us by radio, and wished us good luck.

Shortly after our return to China the snarl of governmental red tape was cleared, and the Shanghai-Canton air line was inaugurated.

A novel feature of our equipment was the floating stations at each stop. They were built of bamboo, as lumber is very scarce in China.

Eighty percent of our passengers were Chinese. They are poor sailors and prefer to travel by air rather than by the small coastwise steamers. Furthermore, Chinese steamers are in constant danger of being captured and looted by the pirates on the coast.

After nearly a year in China I was recalled to New York to attend a conference about additional equipment and personnel for the Orient.

CHAPTER TWO

WHEN I left China I expected to spend about two weeks in New York and then catch the next steamer back. But upon reporting to headquarters it was made clear that I might not go back at all. More important plans were brewing. After much discussion it was decided that I would return with the additional equipment and personnel for China as soon as they were ready, but must hold myself in readiness to join the transpacific division on short notice.

I had two months to kill before returning to the Orient. The Chief Engineer sent me to Bridgeport, Connecticut, where the first Sikorsky Clipper, the S42, was now ready for preliminary test flights. I spent a week studying several S42's in various stages of construction. The S42 is a sturdy flying boat, built of metal throughout. It has clean lines and is very fast.

Colonel Lindbergh and Captain Edwin Musick, ranking Pan American pilot, were in charge of the S42 flights, and I joined their test group. It was immediately apparent that this plane had a better performance than any flying boat ever built before. So Lindbergh and Musick arranged for official observers and proceeded to break most of the world's seaplane records. They flew the ship for eight hours over a measured course and averaged one hundred and fifty-seven miles per hour, using only seventy percent of the power available. When fitted with standard equipment, the plane, which had been designed for the South American trade, had a range of eighteen hundred miles. This was not enough for the twenty-four-hundred-mile flight from California to Honolulu. But when fitted with extra gas tanks it could easily make this distance.

After several weeks at the Sikorsky factory I visited the Martin plant at Baltimore. There the *China, Hawaiian* and *Philippine Clippers* were under construction. These sister ships were designed for transoceanic work and were ten thousand pounds heavier than the Sikorsky S42. Their wing-spread was one hundred and thirty feet and from nose to tail they measured ninety-two feet. Fully loaded each would weigh fifty-one thousand pounds. The specifications called for an extreme range of almost four thousand miles. The first plane, the *China Clipper,* would not be ready for a flight test for many months, but the important features of the design, and the methods used by the Martin engineers to save weight without sacrificing strength, could easily be seen.

The *China Clipper* would carry a load greater than its own weight. Other manufacturers had been frankly skeptical that such a performance could be secured. When Pan American first submitted the requirements for their transoceanic planes, the leading aircraft designers of the country shook their heads and said it couldn't be done. Only Sikorsky and Martin had been willing to attempt such a design. Sikorsky's first plane had already exceeded its required performance.

After several days spent at the Martin plant, I felt certain that the *China Clipper* would be fully as successful as the Sikorsky S42. I went back to Bridgeport.

The Sikorsky S42 passed its acceptance tests and was named *Pan American Clipper*. Captain Musick and I flew it from Bridgeport down the eastern coast to Miami, Florida, where it was to undergo further service tests.

Miami is said to be the only American seaport where more passengers arrive by air than by steamer. Pan American's main seaplane base and operating headquarters for their South American system is located there. The base contains extensive shops and laboratories. An interesting feature of the work there is the school for pilots.

This school was organized to train the senior pilots for transoceanic work. Later it was expanded to include a complete course of training for young pilots entering the company's service.

The course for the senior pilots embraces celestial navigation and advanced seamanship. It is comparable to the requirements for papers as master of an ocean liner. Those who complete the course with the highest marks, are chosen to command the transoceanic planes.

Young pilots who secure employment with Pan American must be graduates of a good technical school. They must have a transport pilot's license and a certain number of hours in the air. These youngsters are classed as apprentice pilots. They are required to spend sufficient time in all departments to become proficient in each. They must qualify for a mechanic's license and a radio operator's license. Later, when they are assigned to duty as junior officers on a clipper plane, they must be qualified to take their turn at flying, as radio operator or as engineer.

Interesting experiments to develop a long range radio direction-finder for us on the transpacific run were in progress at Miami.

After ten days there I headed for Los Angeles, where I was to pick up the new planes for China. In order to brush up on Pan American's latest operating methods, I took a round-about route to Los Angeles by way of our lines, through Panama and Mexico City.

Upon arriving at Los Angeles I found the planes and spare parts had been delivered. Everything was packed aboard ship and we were ready to sail. At the gang-plank I received a wire from our New York office instructing me to cancel my passage and stand by for orders.

I knew this meant the transpacific show was about to open. I had no regrets about China. I'd been away from the States for five years, and it was grand to be home for a while where you didn't have to boil every drop of water, and where it wasn't suicide to eat fresh vegetables.

In a few days orders arrived by mail. They directed me to make my headquarters in San Francisco and begin a survey of the Bay area with regard to weather, flying conditions and airport sites. I went to work and sent many long reports to New York but they apparently were too busy to answer. I was beginning to think they had completely forgotten the Transpacific line. Then one morning at 6 A.M. they telephoned me to take over Alameda Airport in the name of Pan American Airways.

Alameda Airport is directly across the bay from San Francisco. It borders the bay, and has a yacht harbor on the outboard side. This yacht harbor would provide quiet water, deep enough to permit handling and beaching the clipper ships. The airport had an administration building, a large hangar at each end of the field, and a row of small hangars.

The buildings had been neglected. I hired a few men and we started to clean house.

One morning I received a call from one Theodore Ching, who had worked for us in Canton, China.

"Ching," I said, "what are you doing over here?"

"Well, Mr. Grooch, when you didn't come back to China I decided I'd come over and go to work for you here." Ching's mother is Chinese and he speaks Cantonese. He had been very useful to me in Canton on one occasion when we had the mayor and his staff down to christen our first mail plane. None of them spoke a word of English, and Ching served as interpreter. Ching is an American citizen and a good mechanic, so I put him to work.

Shortly thereafter company executives, personnel and equipment began to arrive, and Alameda Airport soon

SKYWAY TO ASIA

hummed like a twin-row Wasp engine. The *Pan American Clipper* was due before long and we had to be ready to receive her. The only hangar at all suitable for our equipment was too low to take the clippers and had to be raised twelve feet. Our contractor put house jacks under the roof trusses, unbolted the feet of the columns and jacked the entire hangar twelve feet into the air; then he riveted extensions to the feet of the columns, bolted the extensions to the foundations and the hangar was ready for use. This was a strange-looking operation but it was entirely successful.

We removed a section of the concrete breakwater at the inboard side of the yacht harbor, and constructed a wooden carpet ramp on which to haul out the planes on their cradles. Then we built a hard surfaced roadway from the head of the ramp to the hangar.

The Chief Engineer directed me by letter to let bids for the construction of two large sea-going motor launches and three large lighters, and to supervise their construction. The launches were to be used as tow boats for rescue work at the mid-Pacific island stations. The lighters were to be cargo carriers at the same points.

Frank McKenzie, formerly an engineer for Curtis Airport at Alameda, had been employed by our New York office and was detailed to the Alameda Airport at this time. Mac was a great help as he knew where to turn for anything we needed in the Bay area. He also let bids for much equipment. Both of us were very busy.

This was all hard work and good fun but it was still difficult to realize that it had any direct bearing on a Transpacific airline. The flurry of activity at the airport increased from day to day.

Then orders for the "North Haven Expedition" arrived.

CHAPTER THREE

PAN AMERICAN AIRWAYS had developed planes that could fly across the Pacific Ocean if arrangements could be made for them to stop and refuel at regular intervals. That "if" stood for a big job. It meant that we had to build air bases in the middle of the Pacific.

The purpose of the North Haven Expedition was to construct those bases. The route decided upon led from California through Honolulu, Midway Islands, Wake Island, Guam and Manila to Macao, China.

The only way to get to those places was in a ship. It had to be our own ship as some of those islands are far off the steamer lanes. So our company had chartered a ship, the S.S. *North Haven,* which so far sounds simple. All we had to do was pile our materials on the ship, send the ship to the islands, unload the cargo and hire a local contractor to construct the stations. It wasn't a bit like that.

In the first place, two of those islands are pin points on the chart and it takes a good navigator even to find them. They are entirely surrounded by coral reefs and the ship must unload her cargo in the open sea. There are no local contractors, no workmen, not even fresh water. The ship would have to furnish every single item necessary to construct the stations, and stand by outside the reef until the stations were in condition to maintain themselves.

The newspapers dubbed the North Haven Expedition the "Modern Robinson Crusoes," but they overlooked the fact that that gentleman had no idea of inhabiting a desert island and would never even have seen the place if he hadn't been shipwrecked; while our expedition knew just where it was going. At that, Crusoe picked a better island than we did because he at least had fresh water and fruit. We might have picked better islands except for the fact that good islands are very scarce in those parts. Wake Island is twelve hundred miles away from any other island. Its charm for us lay in the fact that it was right on the route to be flown, was spaced right between Midway and Guam, and had a nice lagoon for the transpacific plane to land in.

The plans for the expedition were forwarded from New York as rapidly as they were completed. They had kept a corps of engineers busy for months.

The New York office selected me for the expedition because I had had marine experience and also understood aviation requirements. I established my headquarters on Pier 22 in San Francisco. All materials were addressed to that pier, and were to be loaded on the *North Haven* when she arrived a month or so later.

My first move was to look for an assistant to check the cargo as it arrived. The dock superintendent recommended one Dan Vucetich for the job, and Dan came to see me.

"Mr. Vucetich, are you an American citizen?" I asked.

"Yes, Mr. Grooch."

"The dock superintendent tells me you speak several languages."

"Yes, sir. My father took me with him on his travels when I was a boy, and languages are easy for me. We traveled in Europe for several years, and I came to know nine languages."

"What sort of business experience have you had?"

"Mostly banking," he said, "but I am familiar with shipping procedures, and I know how to check cargo."

"When can you report for duty?"

"Right now, sir."

"Very well, you're working for Pan American. Let's have a look at things on the pier."

Dan took hold right away and handled his job in fine style. He had plenty of worries but always kept his temper in hand. If he did have to swear he swore in Turkish and so avoided hurting anyone's feelings.

We looked over the pier together and made plans for stacking the materials as they arrived. The pier was eight hundred feet long and could accommodate several ships at once. Dan and I thought half of the space available would be ample for our needs. We were marking out and labeling spaces where certain materials were to be stacked, when Al Mittag arrived from New York. Mittag was a young engineer who had been working on plans for the midpacific bases, and had helped order the materials.

He shook hands and asked if I had received a master shipping list of our cargo.

"Not yet," I replied. "The purchasing department has promised to send me a list as soon as they can get one completed."

"Well here is one almost complete," said Al. "It will give you an idea of the amount of stuff you've got to receive, separate and stack."

"How are they shipping this gear, Al, and where is it coming from?"

"Some of it will come from the east coast by rail in carload lots, and broken shipments from a dozen different states," he answered. "Late orders will come by rail express."

We studied over the master list together and decided that we would need all the space on the pier to store our cargo. The concrete floor was divided into squares by heavy chalk lines. The area of each space was decided by the amount of the commodity to be stacked there. Large signs were hung on the wall to indicate what goods each space should contain.

The material came in slowly at first. One day an irritated shipper called me up. "Where the hell do I send this carload of windmills?" he asked.

"Send them right down to Pier 22," I replied.

"Well I got a lot more queer stuff coming for you," he said. "What's going on down there anyway?"

"Oh, we're starting a couple of seaplane farms out in the Pacific," I answered. "What else have you beside windmills?"

"I've got some big crates marked 'Solar Heaters' and a lot of power-house equipment," he said.

"Send 'em over. Hereafter send anything over as soon as enough arrives to make a truck-load."

After a few days every steamer and freight train that arrived in San Francisco handed us cargo faster than we could handle it: ten-ton tractors, five-ton generators, power-house equipment, radio gear, refueling equipment, dock timbers, fifty-foot radio poles, mountains of furniture, stoves and household equipment, complete cold-storage plants.

Our greatest concern was to segregate and pile this equipment in such fashion that when the *North Haven* arrived we could load it aboard properly. We had cargo for Manila, Guam, Wake, Midway and Honolulu. It must be stowed on the ship in that order so that when we arrived at any station the cargo for that point would be on top and easily accessible.

The dock superintendent was not in complete sympathy with our ideas in the beginning, and thought we were entirely too fussy about our cargo. The dock hands had a marked tendency to dump their loads on the first pile they came to, and we had to watch closely to avoid this earnest attempt to mix up our plans. They finally saw the light.

Our chief airport engineer, Captain Odell, arrived from New York about the middle of March to supervise the final arrangements for the Expedition. About the same time his assistant, Charlie Russel, arrived with two cub engineers named Taylor and Borger, Dr. Kenler who was to be stationed at Wake Island as medical officer, and Mr. Ward who was to have charge of the commissary departments at both Midway and Wake.

McKenzie and I had already picked out a number of good men for the construction crew. The news got around, and each day more and more applicants besieged the airport and Pier 22. Some of these men had good jobs but the expedition promised excitement and adventure, and they were sure to save some money as they couldn't spend a cent where we were going.

Dr. Kenler gave the chosen men a physical examination, vaccination and typhoid inoculations.

Ground crews for the midpacific stations were selected from Pan American's Caribbean or Mexican divisions, and were all men of long training in aviation. These men began to roll in from all directions. As soon as they reported, we started to instruct them in their new duties.

The airport managers were assigned to each department at Alameda Airport for several days, to get a general idea of how the island bases should be operated. Chief me-

chanics and assistant mechanics reported to the shop superintendent for indoctrination in the plane service routine
to be followed.

Each air base was to have a radio station and the new-
est type of radio direction-finder. The radio ground
crews were recruited partly from our other divisions and
partly from local sources. They reported to the communication superintendent for instruction in radio procedure, and in operating the radio direction-finder.

Our first taste of sea-going life came when we tested
out the big new motor launches that the Kneass Boat
Company had built for us. We gave them a speed run in
the bay and then took them out to sea to note how they
handled in rough water. They were not fast, only twelve
knots, but they were good sea boats and fine for towing.
As rescue boats they might have to go five hundred miles
to sea, so they had a range of a thousand miles. They
were very sturdily built.

On March 22, 1935, the *North Haven* arrived in San
Francisco.

CHAPTER FOUR

THE *North Haven,* our chartered steamer, was a well found oil-burning vessel of some five thousand tons. Her home port was Seattle. Before leaving there she had taken aboard the materials for the buildings we were to construct at Midway and Wake Islands. These materials had been cut and furnished by a Seattle mill which sent along a construction man to show us how they went together.

Soon after the *North Haven* docked I went aboard to interview the captain. The first person I ran into on deck was the steward, Fred Scott. Scott was a little Englishman of half-pint size who had been to sea most of his life. He took me to the skipper, Captain Borklund. The captain was a North Country man from Sweden or Norway, a naturalized American, and a dependable looking chap with a kindly manner. He rose from his desk to shake hands.

"Goot mornin'," he said. "I vas yoost going down der dock to der telephone. De owners dey say I should report to you so soon as I come."

"Glad to see you, Captain," I answered. "You've got a staunch-looking ship here. What sort of crew did you sign on?"

"Vell some of dem could pe petter but dey vas de bess I could get."

"Is your deck-gear and hoisting tackle all in good shape? We've a lot of heavy stuff to handle."

"Everything is fine, I tink," he replied.

"Have you got a good first mate?"

"Mr. Petersen iss a goot man."

"Good. Now let's have an understanding on your duties and mine. You are captain of the ship and will be responsible for its navigation and for the conduct of your crew. I represent the charter party and will direct you as to the movements of the ship and be responsible for the conduct of my men. Does that check with your orders?"

"Yoost vot de owners tell me," the captain agreed.

Formalities complied with, the captain and I sat down to an hour's conference on the difficulties of the job ahead of us and the methods to be used. We called in the first mate and discussed the proper stowage of cargo. Five thousand drums of aviation gasoline had to be stowed in the lower holds, along with a lot of Diesel oil and lubricants. I requested the captain to take on bunker oil immediately and then proceed across the bay to the Standard Oil docks at Richmond and load the gasoline.

The *North Haven* was loading this gas at Richmond when the tanker strike descended upon San Francisco. Standard Oil operated tankers and was therefore on the black list, and no union stevedore or seaman could touch Standard Oil product. The strike committee ordered the crew of the *North Haven* to walk off the ship but

[28]

Captain Borklund persuaded them to allow the crew to remain, by promising to leave the Standard Oil docks at once and take the ship back to San Francisco, which he did.

The situation looked bad for us as the strike appeared to affect all the oil companies. There were strong indications of a general waterfront strike which would tie up all shipping. Time to us was precious. The Standard Oil officials could offer no solution. Barely started, it looked as though our expedition were on the rocks, with a chartered ship on our hands and a large construction crew under contract.

By accident I discovered that the Shell Oil Company did not operate tankers and had not yet been placed on the black list. We asked the strike committee if we could load Shell products. They gave us permission but stated that Shell Oil might go on the black list any day.

Fortunately the *North Haven* had loaded sixteen hundred drums of Standard aviation gas before she had been ordered to stop. If we could load a few hundred drums of Diesel oil and lubricants before the strike committee changed their minds, we could carry on. We hired fleets of trucks to transport the desired material from the Shell Oil plant forty miles away. They ran day and night until the stuff was loaded. We breathed a sigh of relief when that hurdle was behind us.

Meanwhile the mass of cargo on the pier was being trucked and swung into the *North Haven*'s holds. Dan

Vucetich and half a dozen checkers were on the job twenty-four hours a day, endeavoring to make sure the cargo was stored according to our plan.

On the morning of March 26 we were certain that barring unforeseen trouble we would be ready to sail before midnight. The construction crew was ordered to assemble at the airport that afternoon prepared to go on board.

We held the men at the airport until just before sailing time and then sent them over in trucks.

The newspapers printed a story stating that the North Haven Expedition was surrounded with a veil of secrecy. This was inspired by our endeavor to avoid damage to our cargo.

Several hours before sailing time a crowd began to assemble in front of the pier, which was the last thing we desired. Friends, wives and sweethearts were passed through the lines but the crowd of onlookers were held back until the gangplank had been hauled aboard.

I came down in a taxi with several friends about two hours before sailing time, having sent my luggage aboard some hours before. Because I had been constantly on the pier for several weeks, I knew all the regular dock police. But this last night there were some new boys on guard who didn't know any of our party. Their orders were to let no one through without a pass. Although I had signed all the passes I had forgotten to supply myself with one.

I approached the policeman on guard. "Sergeant," I

said, "I'm in charge of the Pan American party on the *North Haven* and I've got some friends with me that want to see us off."

"Have they got a pass?" he asked.

"No, but I'll vouch for them."

"Well, let's see *your* pass."

"I left it in my room on the ship. If you'll send one of your men with me I'll show it to him."

The cop smiled. "Buddy, you're the fifty-first guy that's tried that stall tonight. Don't waste my time."

From the nearby crowd came a number of helpful suggestions.

"Stay with him, boy."

"Tell him you're President Roosevelt."

"Why don't you sock him?"

At this moment, to my surprise, the taxi driver stepped up to the policeman and said, "I know this guy, Joe, and he's O. K."

The cop was annoyed. "I don't care who knows him," he said. "He's gotta have a pass to get through this gate."

Fortunately the dock superintendent came along and persuaded the cop to let us through.

The chauffeur drove us down the pier to the gangplank. While paying him off I noticed that his face seemed vaguely familiar.

"Haven't I seen you before?" I asked.

The driver grinned broadly. "Sure you have. I was a flight mechanic with your outfit in Brazil. Don't you remember the revolution of 1930 and that night in Pará when we were barricaded in the Grand Hotel?"

I recalled the man at once. "Of course I remember," I said. "How are you, Jenkins?"

"Just fair, Mr. Grooch. Don't make much money but the wife's folks live out here and she likes it better than South America."

"Why don't you come along on the *North Haven*?"

"Oh, we got a couple of kids. The old lady'd squawk her head off."

"Well, good luck."

"The same to you," said the driver. "Looks like you'll need it."

Our construction crew had been mustered at the gang-plank. Several of the men had been too busy celebrating their impending departure to show up. McKenzie knew that fifty or more men at the gate wanted to go. He signed up three of them. Two lived near by and rushed home for suitcases, but one chap lived too far away to go home and get back in time so he came aboard "as is." His only baggage was a ukulele.

The Pan American public-relations man, Van Dusen, came aboard with Junius Wood, a well-known press correspondent, in tow, and introduced him to our party. Junius was going along with us to write the story of our expedition for the newspapers.

Shortly before midnight the last package was aboard. A final muster was taken to insure that none of our party was being left behind. The whistle bellowed hoarsely, lines were cast off and the *North Haven* backed away from the dock amid a clamor of shouted farewells between those on board and those on the pier.

After leaving the pier we anchored for the night in San Francisco Bay, as a radio compass, which had just been installed on the ship, had to be tested and calibrated before we left port. This operation had to be performed away from docks and obstructions. The next morning at 7 A.M. the R.C.A. electricians came aboard and completed this job with dispatch.

The morning papers spoke of a "mystery ship that went out through the night." Actually we weighed anchor at 10:10 in the morning. We stood out through the Golden Gate, dropped the pilot at 11:40 and shaped a course for Honolulu.

Shortly thereafter a red airplane zoomed over us. In the cockpit was our friend, Mike Doolin, manager of the San Francisco Airport. Mike learned how to stunt a plane in France, and he proceeded to show us that he had not lost the knack. After giving us a good show he flew low alongside, waved goodbye, and zoomed away as we steamed west.

CHAPTER FIVE

ON our first day out there was a stiff breeze from the northwest and a rough beam sea. Nearly all of our crew were landlubbers and most of them failed to show up for the noon meal. The wardroom mess took to their bunks in a body. Some of them nearly starved to death before they could hold any food down. The little steward, Scott, was greatly concerned and fluttered from room to room offering to prepare any delicacy that anyone thought he could eat, but there were no takers. Most of the crowd lived on orange juice for several days.

The ship rolled so badly that it was necessary to shift course to ease the roll and to put extra lashings on the deck cargo. During the first two or three days at sea those who were able to get about were occupied in trying to acquire a pair of sea legs, to find out who was who in the party, and what it was all about.

The captain and the ship's officers all had a strong brogue, but they were friendly, accommodating, and good sailormen.

Our wardroom crowd was pretty well sandwiched in because of the scarcity of space. McKenzie, Mittag, Borger and Taylor shared a four-place cabin. All were seasick and they had a sad time for the first several days.

Russel, the senior construction engineer, had a cabin to himself but that was small comfort to him as he was plainly in doubt as to whether he would live to reach Honolulu.

Junius Wood shared a small outside cabin with John Steele, our service engineer from the New York office. Both of them had been on sea trips before and were fair sailors.

Down amidships the crew had ample room and soon settled down comfortably. Few of them knew any of the others to start with, but they quickly got acquainted. Quite a few had been out of work, and as a consequence had been dieting. The ship served excellent meals, so they made a noble attempt to take advantage of their opportunities regardless of their seasickness. This heroic attitude was rewarded and most of them were soon up and about.

As always happens in any group of men, the ones with qualities of leadership were easily recognized. Among this group were George Kuhn, boat carpenter; George Robinson, boss plumber; George Ferris, master boatman, and Bob Kratz, boss electrician.

Our crew were mostly landsmen and of course knew practically nothing of seamanship. We couldn't afford to take along a lot of sailors because sailors would be no good on our construction jobs. Yet we had a lot of seagoing tasks before us. So, having no sailors, we proceeded to teach our landlubbers all the sailor tricks we knew, and had them learn how to splice rope, tie knots, and make up cargo nets, fenders and tow lines.

In view of the strenuous nature of the work ahead, we instituted semi-military discipline. One chap had a slide trombone. He was promptly elected company bugler. He took particular pride and delight in blowing *reveille*. Every man was allowed fifteen minutes to make up his bunk and be ready for breakfast. After breakfast there were illustrated lectures on the details of the work to be done at the islands in order to familiarize each man with his job.

The first mate, Petersen, was vastly interested in our program and gave us many excellent pointers and much good advice.

It was interesting to note, however, that there was absolutely no mixing between the ship's crew and our construction gang. The ship's crew were union seamen and had no dealings with anyone who didn't belong to their union. The ship paid them sixty dollars a month and found to perform their ship's duties. But if they so much as touched our cargo or removed a hatch cover for us we had to pay them sixty cents an hour and for overtime seventy-five cents an hour.

In our wardroom mess we had two young men who were going out to be the managers of the airbases at Midway and Wake Islands. Leuder, the manager for Midway, was small, serious and soft spoken, while Bicknell, the manager for Wake, was six feet four and weighed two hundred and thirty pounds. Part of the duties of these men would be to put to sea in our big launches in case it were ever necessary to go to the rescue of a plane forced down. This required a knowledge of navigation and our cruise offered them a good opportunity to learn it.

Fortunately I was competent to serve as instructor as I had been required to know navigation thoroughly in the Naval service. So a school was started. Each morning, noon and night there was a parade of embryonic navigators to the bridge to shoot the sun, moon or stars. Many facetious comments were offered by members of our party as to the results of our efforts, and the *Scuttle-butt News*, a libelous sheet edited daily by our crew, hinted slyly that there was a broad disagreement among the navigators as to whether we were in the Atlantic or the Pacific.

By and by, however, our students began to get the idea. Each day at 12:30 the second mate would hand me a memo showing the captain's position at local noon. As soon as our students began to feel sure of their results, they took great pride in arriving in the wardroom just before the mate, each with his own position worked out. Usually they would be pretty close to the captain's position and loudly insisted on all hands taking note of the fact.

After much discussion of the problems to be met at Midway and Wake, we decided that it would require more men than we had on hand to finish the work within the time limit set forth in the ship's charter. We therefore decided to hire some men at Honolulu in addition to the

Chinese domestic servants we expected to pick up there.

As we neared Honolulu the weather grew warmer, skies were blue, flying fish skimmed over the sparkling water, and sunbaths were the order of the day. Our crew resembled a nudist colony, and grew more tanned and more salty by the hour. Their talk was all of ships and sailormen. Every man of them soon believed that his ancestors had roved the Spanish Main and was confident that he could tame any island in the Pacific with one hand tied behind him.

We rounded the big dark promontory of Diamond Head and sighted Honolulu shortly before noon. At quarantine we were met by the port authorities, our local manager, and newspapermen.

It was immediately apparent that the people of the Hawaiian Islands were deeply interested in our endeavor to develop a transpacific airline. There was a charming lack of formality and a whole-hearted desire to help that made us feel at once they had faith in us, and wanted us to succeed.

We spent three days in Honolulu. A few of our party were very busy, but a large part of the crew had leisure to look around and visit points of interest. They fell in love with the place.

I had been hoping I might be able to dig up some firsthand information in Honolulu about Midway and Wake. Fortune favored me. I ran across a man who had recently been to Midway Islands for the Commercial Pacific Cable

Company, and I also talked with a Naval aviator whose ship had lately stopped at Wake Island. Neither report was at all reassuring but at least they gave me a fair idea of what to expect.

It appeared that we would have to do quite a bit of blasting to clear coral heads and construct channels for boats and planes. Acting on this information we purchased a ton of dynamite and some drill tools, heavy hobnailed shoes for all hands, and sun helmets.

As none of our party had had any experience in blasting coral, it was necessary to pick up a good powder man. Several dredging firms recommended one Dave Richards, a Kanaka. Kit Carson of the Matson Steamship line called up Dave and told him to meet us at Carson's office.

Dave arrived with a couple of leis around his neck. He was in a very expansive mood.

"Dave," said Carson, "Mr. Grooch wants a good powder man for the North Haven Expedition. How about it?"

Dave turned to me. "What you goin' to shoot, coral?"

"I don't know just what's ahead of us but the chart shows coral heads in the lagoons at Midway and Wake Islands. We may also want to clear a boat channel through the reefs."

"How much powder you got?" he asked.

"We're loading a ton of dynamite tomorrow. Do you want to come along?"

"That ain't much powder," said Dave, "but I'll come."

"O. K., but get this straight. None of our crew knows a thing about handling powder. I want you to take charge of the stuff and be almighty careful with it."

Dave promptly waxed loquacious. "I been a powder man for twenty-five years. If you got some shootin' to do just gimme enough powder — that's all. Folks thinks powder is bad stuff. Powder ain't dangerous by itself; less'n you put a cap on it you kaint hardly make it go off. One day on Molokai we dropped a case off the sling and it fell eighteen feet on the rocks. It didn't go off. And one day . . ."

"Hold on, Dave," I broke in. "Remember you're going out with a green crew. That stuff you're telling me is bad medicine for them. If you want to keep your health, handle your powder with kid gloves and make your helpers do the same."

Dave turned out to be a careful and expert powder man. He was around sixty years old. Being a typical Kanaka, he had more native superstitions than a clipper's instrument board has dials, but he also had the sunny temper and the carefree disposition of the South Sea Islander, and was soon a prime favorite with the entire crew.

We let the word get around that we wished to hire a dozen or so more men for the construction crew. As soon as this news leaked out, half of Honolulu volunteered to go along. Among the candidates were several seamen. They were likely looking lads and we signed them up at once. After leaving Honolulu we discovered they had jumped ship from a steamer in port. One of them had second mate's papers. When I asked him what the idea was he said, "Well, sir, we wanted to see the fun."

Later we learned that their skipper had made a big fuss in our office in Honolulu, claiming I had enticed them away from their ship, but we were halfway to Midway by that time.

We also needed a dozen Chinese boys for domestic servant jobs at Midway and Wake. Honolulu had many Chinese willing to go with us, but the business of hiring them turned out to be an intricate affair.

The Chinese get along peacefully enough with people of other nationalities but disagree scandalously among themselves. They have their own brotherhoods or tongs and if several boys from different tongs are carelessly mixed, you may expect fireworks any time thereafter. In the effort to select a clear strain, we visited several prominent Chinese gentlemen and solicited their advice and assistance.

They recommended a number of candidates and we picked out a baker's dozen. While we were interviewing them a police official strolled down our way. Our candidates betrayed marked uneasiness and several took to their heels. The police official good-humoredly explained that most of our candidates were the bad boys of Honolulu —hatchet men and gangsters. Departure time was drawing near and we had to have camp cooks and waiters.

We finally managed to sign up a dozen Chinese who passed muster with the police. Each of them insisted on a cash advance to outfit himself for the trip. I advanced the money with strong doubts as to whether I'd ever see any of those Chinese boys again. But much to my surprise they all reported aboard well before sailing time.

Assisted by a number of the citizens of Honolulu, we got our affairs in order for departure. The dynamite was placed aboard with much ceremony, as, thanks to Dave, our crew had a wholesome respect for it. We wrapped it in swaddling clothes and parked it in a forward compartment under lock and key.

No one ever leaves Honolulu with a grouch if the local boys can help it. Our departure was distinctly a cheerful spectacle—every man with a lei around his neck, and a friend to wish him luck, the band playing "Aloha," and Dave Richards dashing around assuring all hands that he'd radio the time of our return and begging them to have a hundred hula dancers standing by.

We backed away from the dock at 5:30 P.M., each man vowing that he'd see that town once more before he died. The *North Haven* pointed her nose for Midway Islands, thirteen hundred miles away, and the long Pacific swell began to roll under our starboard quarter.

Midway Islands derive their name from the fact that they are about half way from California to China. They are the last of a chain reaching out from Honolulu slightly north of west. Our course lay some twenty-five miles to the north of this chain, and we saw nothing of land except occasionally the dim outline of a high hill on the horizon.

SKYWAY TO ASIA

One day the lookout sang out, "Shoal water ahead!"
The captain jumped for the bridge with me close behind
him.   Sure enough, a mile or so ahead, was a patch of
white water looking exactly like a shoal with swells breaking over it.   The captain turned to the man at the wheel
and said, "Give her two points to starboard."

We stepped into the chart room and took a quick look
at the chart.   It showed nothing but deep water near our
position.   The captain grabbed his binoculars and went out
on the bridge.   After looking for a minute he grinned
and handed the binoculars to me.   As soon as I focused
the glasses on the white patch ahead, I also grinned.   The
shoal was an enormous school of dolphins.   Thousands of
them were at play, leaping out of the water, throwing
up clouds of spray.   The captain put the ship back on her
course.

The days passed quickly.   We were busy with navigation and our final preparations for the assault on Midway
Islands.   On the morning of April 12, all hands were on
deck before daylight to catch the first glimpse of the low-
lying land, which soon appeared out of the sea ahead.   We
anchored off the southeast reef at 7 A.M.

CHAPTER SIX

OUR anchorage was half a mile from the circular reef that completely surrounds Midway Islands. From the bridge we could see a great circle of foam caused by the breakers pounding on the shallow coral. The circle was five or six miles in diameter; on its south side lay two small islands about a mile apart. Over both of them clouds of sea birds were wheeling, but there were no other signs of life. We could hear the muffled roar of the breakers and through the glasses could see that the far side of the lagoon was the weather side, as there the foam and spray were thrown high into the air as the big swells dashed against the reef. We could see no passage through the circle of breakers.

We knew that there was a cable station on one of the islands, and presently our radio operator reported that he had picked up their radio station but that its signals were very weak. I sent a message asking if they had a launch, and if so could they come out and show us how to get through the reef. Shortly afterwards I received an answer from Mr. Perry, the superintendent of the cable station, stating that he would come out in his launch through the southeast passage. By studying the chart I found that we were anchored off the southeast passage but neither the captain nor I could see anything but an unbroken line of white breakers ahead.

Our crew manned the rail and listened to the far-off music of the sea. Every man had a serious look on his face. Presently a launch rounded the shoulder of Sand Island, the larger of the two. The launch crossed the lagoon, approached the inside of the reef, and then lay to for several minutes, apparently awaiting a favorable moment to make the crossing. It looked like suicide from where we stood and we hoped the men in the launch knew what they were doing. Suddenly the launch moved forward and was immediately hidden from view by foam and spray. We were sure they had capsized, but suddenly they emerged from the line of foam and headed for the ship.

They came alongside the gangway and Mr. Perry and his engineer stepped aboard while the boat crew hauled out and tied up to a line astern. All of them were a bit damp from the spray. Mr. Perry and his engineer, Mr. Lawn, were met with a genuine welcome. They were fine chaps, but we would have welcomed the devil himself if he could lead us safely through those reefs.

We sat around over morning coffee and deluged our visitors with a flood of questions. They had been informed by cable that the *North Haven* would arrive sometime that week, but the daily news flashes had given them only a vague idea of the purpose of the expedition. They received mail only three or four times a year, brought to them by the cable company's supply ship. When they learned that Pan American would soon inaugurate a weekly airmail and passenger schedule, through Midway, they were, to say the least, astonished. It developed that Mr. Perry had been at Midway for only a short time, but Mr. Lawn had been there for six years, and knew the reefs like a book. The latter was doubtful that our big launches could navigate the narrow channel through which his boat had just passed. So I took him out on deck to inspect our craft.

I asked him about the width and depth of the passage.

"It's about twenty feet wide," he answered, "but it isn't straight. There's a bend at the center. There's about five feet of water over the reef there now, but when a swell runs out from under you there's a few seconds when there's only three feet."

"Our launches draw four and a half feet," I said. "We can't afford to risk losing them in that passage. Is there another channel?"

"Yes, there's a much better channel on the northwest side. It's all right when the swells aren't breaking. But that's the weather side and you'll have to watch your step."

"Do you think we could get through there today?" I asked.

"Well, the sea is calming down a bit. By this afternoon we ought to be able to make it."

"That's fine. Could you pilot us in and have your launch follow?"

"Why sure."

Mr. Perry agreed to this. That afternoon we prepared to lower one of the big launches.

The launch weighed fifteen tons and was thirty-eight feet in length. The business of lowering a boat of this size at sea requires careful preparations and excellent seamanship. First Mate Petersen had charge of the lowering. He knew exactly what to do and his orders were definite and precise.

The boat was lowered over the side with only the boat crew aboard, the ship's crew manning steady lines. When the boat was within a few feet of the water the mate held her for a couple of minutes while he watched the roll of the ship. At just the right moment, when the ship was steady, he gave the order to lower rapidly and the boat entered the water with hardly a splash.

The boat then made a short run to test the engines, and came alongside the gangway.

Mr. Perry had invited most of our wardroom mess to come ashore for dinner and spend the night, so we boarded our launch and set out for the northwest passage with their boat following us. Mr. Lawn acted as pilot. He and I stood on the top of the pilot house in order to get a clear view. As we circled the reef the swells increased in size until they were running ten or twelve feet in height. One minute we were deep in the trough, and the next we were high on the crest of a swell.

Our launch was a good sea boat and climbed those advancing walls of water with little effort. After several miles of this we arrived off the northwest passage and laid to for a moment to see if the swells were breaking.

The passage was merely a low space in the reef about a quarter of a mile in width. Mr. Lawn watched closely while a dozen swells rushed through it. He told us there were eighteen feet of water over the reef in the passage but when the swells were running high they often built up as they crossed the reef and turned into breakers easily capable of capsizing our launch. On our left was a broad section of the reef known as "North Breaker," invisible due to the mass of foam and spray which covered it. On our right was a curved section of the reef known as "The Hook." This was also invisible, with a wicked looking swirl of foam and spray to mark its presence.

The swells were not breaking in the passage so we started through. The launch operator opened his throttles wide and we held full speed in order to have good steerageway. The big swells rushed down on us from astern, lifted us high and then ran out from under, dropping the launch deep into the trough. The tops were trembling and leaning forward, just ready to fall. Another six inches in height and they would have been real breakers. As each swell overtook us, it seemed inevitable that the towering mass of water would overwhelm the launch. But each time the stern of the launch lifted skyward and then pointed down-hill as the swell passed. The effect was something like a rollercoaster.

As we advanced into the lagoon it was seen to be slashed

and criss-crossed with numerous reefs. Some of these were twenty feet below the surface but the water was so clear they appeared to be barely under it. There were no channel markers. As it was growing dark, Mr. Lawn hesitated to take our deep-draft launch all the way into the beach for fear of running aground. So we moored to a buoy half a mile inside the reef, and proceeded ashore in the other launch.

The entire staff of the cable company was assembled on a small dock in front of their station to greet us. The operators were mostly young fellows, very pleasant as they welcomed us to Midway Islands. Their station was built around a central square covered with a well-kept lawn bordered with flowers. The buildings were mostly of re-inforced concrete, twenty-five years old, but in excellent repair. There were concrete or graveled walks, many trees and shrubs, and the general effect was peaceful and lovely. The station was surrounded by a hedge of trees and shrubbery to shield it from the sand storms which are of frequent occurrence.

Numerous canaries flitted about the compound. They had multiplied from a few originally imported. Beyond the compound was a vegetable garden, pig and chicken runs, and a cow. All of these comfortable surroundings were the result of years of patient labor and experiment by cable-company employees.

When the cable company had first established the station, Midway Island was a barren wind-swept stretch of sand.

Soil was imported for the compound and the vegetable garden, and the wide world was searched for trees and shrubs that would grow in sand. Now the entire island is covered with a shrub known as magnolia brush, and there are several clumps of fair sized trees which were imported from Australia.

The cable-company dinner that night would have done credit to any first-class hotel. We had turkey with cranberry sauce, and a huge cake with "Welcome Pan American" scrolled in the icing. We toasted the occasion with an excellent wine, and Mr. Perry made a formal speech of welcome. For a desert island Midway rated high.

We were up with the birds next morning and made a quick survey of the section of the island where we were to build our station. We followed a path which led from the cable compound to a grove of large trees on the opposite side of the island, and found that this grove was well within the limits of our concession.

The land was fairly level but covered with a dense growth of magnolia brush. It was apparent that merely clearing enough ground for our station would be a big job. The concession was encircled on the outboard side by a serried row of sand dunes forty to fifty feet high. These had been built up by the sand drifting before the wind, and now maintained their shape and location because they had become covered with magnolia brush which held the surfaces together.

Beyond the dunes was a circular stretch of flat sand

[51]

beach fifteen hundred feet wide, only a couple of feet above high tide. On the north side of this beach deep water ran fairly close to shore. We decided that we would land our cargo there. We hurried back to the cable station and radioed the *North Haven* to move over and anchor off the northwest passage.

After months of planning and theorizing, with only vague information to guide us, we were face to face with the facts. The odds for or against us would depend largely on the weather. All of us had the natural reaction of football players before the kickoff — a nervous desire to get the ball into play and get the suspense over.

CHAPTER SEVEN

THAT afternoon the cable company's launch took us out to our launch. We climbed aboard and headed out through the passage in the reef for the *North Haven*, which had moved down to her new anchorage off the North Breaker. The sea had moderated but the swells were still about ten feet high. The ship had to use her engines to warp her head around into the swells before we could put the launch alongside the gangway.

As soon as I got aboard I called a conference, including the captain, the first mate, and the airport engineers, and outlined our immediate plans. Before we could start to tow our cargo into the beach we must first take soundings and mark a safe channel through the maze of inner reefs in the lagoon. We also had to rig a boat-boom aft with a Jacob's ladder in order that we might get aboard the ship when the weather was too rough to permit the launch to lay alongside the gangway.

For the boat-boom we used a fifty-foot radio pole lashed across the stern of the ship, extending some thirty feet horizontally from the side. At the outer end of this pole was suspended the Jacob's ladder, constructed from flexible steel cable with rungs of steel tubing spaced a foot or so apart.

Boat-booms and Jacob's ladders are every-day items in the life of a Navy man, but are seldom used by freight steamers. Our construction crew had never seen such gear. I was hailed as a prophet in a strange country. I eyed the rolling surf and wondered if I would prove with or without honor.

We had a light draft launch on the *North Haven* that drew only thirty inches. We put it overboard loaded with channel markers and sounding gear. This launch led the way into the beach followed by our heavy launch. We felt our way in through the inner reefs, taking soundings, dropping anchored marker buoys here and there to show the way.

The distance from the ship's anchorage to the beach was four miles. The sounding survey took several hours. Time after time we had to back out of a blind alley where the water grew rapidly shallow and led to a dead end on a reef. Patches of coral that apparently were ten feet under the water often turned out to be only a foot below the surface, while others that looked to be on the surface were ten feet deep.

Several good-sized sharks glided about and looked us over. If you've never looked a ten-foot shark in the eye at close range, you've missed a thrill. Every time we took a sounding the sharks nosed down to inspect the lead sinker. They gave every evidence of being very hungry.

We finally completed marking a satisfactory channel to

the sand beach. I transferred to the big launch and ran out to the ship and in again, to prove the channel.

Meanwhile the ship had lowered a lighter and was ready to begin unloading cargo. A ten-ton tractor was the first piece of shore equipment to go over the side. Again the first mate covered himself with glory. It took nice judgment to land that ten-ton tractor on a lighter rising and falling at the ship's side in a ten-foot swell. All hands gathered at the rail to watch the show. Four men were lowered in a cargo net and took position at the four corners of the lighter. Then the heavy tractor was lowered to within a few feet of the water, where it swayed with the roll of the *North Haven*.

Minutes went by while the mate gauged the roll and the rise and fall of the lighter. Should the heavy tractor descend too soon it would crash through the lighter's deck. If too late it would remain dangling in the air. The mate stood with his hands poised over his head with the winchman's eyes glued on them. Suddenly the mate swung his hands down and sang out "down." The winchman let go the brake and slid the heavy tractor onto the lighter with hardly a jar. The next instant the lighter rose high on a swell, but the tractor was aboard and safe.

This was really the opening gun of our construction program. All hands on deck let out a lusty cheer for the mate's seamanship. I once heard Mrs. Hoover say, at a dinner given for Amelia Earhart, that "in America there is only one aristocracy — the aristocracy of achievement."

Judged by this standard there is no doubt that the first mate was number one aristocrat at Midway.

With the tractor we loaded some heavy timbers to be used as an unloading platform on the beach, and two of the big sleds. These sleds were twenty-five feet long and five feet wide, shod with steel. They were to be used for transporting material about the island.

Then the big launch took the lighter in tow and set out for the beach. I went along to act as pilot and to indoctrinate the crew in towing. There is quite a trick to towing a loaded lighter in heavy swells. The secret of success is at all times to keep the slack out of the towline. When a big swell overtakes the lighter it lifts and carries it forward rapidly for a short distance and then runs out from under, and the lighter sinks into the trough. This causes slack in the towline. When the same swell overtakes the launch, it lifts and carries it forward rapidly. If there is much slack in the towline it is taken up with a jerk which is very likely to snap the line or tear out the towing bitts. To prevent this the launch operator keeps an eye on the lighter. As soon as it surges forward on a swell he opens his throttle wide. Then, before the swell reaches the launch, he eases down on the throttle. Our launch operators snapped several towlines, but soon learned the trick. The first load made the beach without mishap.

We beached the lighter on the smooth sand, rigged an unloading platform, and ran the tractor ashore under its own power. The cable-company staff were assembled on the beach. They yelled with approval as the big tractor swung around on the sand and hooked onto the first sled load of cargo.

The next day we began to discharge cargo in earnest. To speed up this work we put over the launch and lighters intended for Wake Island, and divided our crew into several groups. One group remained aboard to assist the ship's crew in loading and handling the lighters alongside. Another group was detailed as a beach crew to unload the lighters when they reached the shore. A third group started to pitch camp in the grove. A fourth group was engaged in making an accurate survey of our concession and laying out the location for the various facilities we were to construct, which included docks, living quarters, windmills, radio stationpower house, cold-storage plant, mess hall, shops and offices.

The weather was very warm and most of our crew was soon bare to the waist. It takes a long period of gradual exposure to make such a practice safe in tropical countries, and our airport engineer, Russel, was worried about the danger of sunburn. He fussed around and exhorted the men to put on their shirts, and also requested Dr. Kenler to warn them. The crew paid little attention to such advice, and as a result there were many sore backs in camp the next day. Russel kept his shirt on but wore English "shorts." He was too busy advising other people to realize his own danger and his legs got so badly sunburned that he was laid up on the ship for several days. After that he hobbled around camp in bandages for a week or so. Everyone kidded him unmercifully, but he was a good sport about it.

We set up regular Army tents in the grove for sleeping quarters, and also a cook tent and mess tent. The cable company agreed to let us have fresh water until we could arrange our own supply, and a plentiful supply of brackish water was secured by sinking a couple of shallow wells behind the kitchen.

The radio party set up an emergency short-wave radio set as soon as they landed. In a very short time we were in communication with Alameda.

The problem of feeding the crew ashore gave trouble at first. It had been planned to send meals in from the ship until we had erected the permanent kitchen and mess hall. This plan was quickly abandoned as the food was cold when it arrived and usually from one to three hours late. So the Chinese cooks and mess boys were sent over, and soon had plenty of good hot food ready on time.

Among the first equipment to be landed was a five-ton electric-generator unit. This was set up on a temporary base and Midway was introduced to its first electric light and its first electric refrigerator. The work of unloading soon was zooming along.

CHAPTER EIGHT

On any large construction job there are certain risks. The most dangerous part of our job was discharging cargo in the open sea, while ship and lighters rolled and plunged in the swells. Many of our men had narrow escapes from being crushed under some heavy piece of equipment as it was lowered onto the lighter. The crews on the lighters were not seamen. They had difficulty keeping their feet while the lighter did its dance — one moment straining away from the ship's side until the securing lines were stretched taut as bow strings, returning on the next swell to dash against the ship's side with a jar hard enough to knock every man off his feet.

Shortly after we started this work one of our men lost his balance and fell flat under a descending panel that weighed two tons. One end of the panel hooked over a keg on the lighter, which saved the man from being crushed. The others in the lighter crew were badly scared but stuck to the job. They pulled the man out from under the panel, discovered that he had only a badly sprained back, and sent him up on deck in a cargo net. The doctor looked him over and put him to bed for several days.

When we started to unload the fifty-foot radio poles

the mate decided to put two of the ship's seamen aboard the lighter because of the unusual hazard involved. The first big pole was swung out over the side and lowered to within a few feet of the lighter where it swayed dizzily as the mate waited for a favorable moment to land it. Watching the gyrations of that pole reminded me of Victor Hugo's famous story of the cannon that broke loose on a ship's deck and charged around like a mad bull.

The mate warned the men on the lighter to stand clear and motioned them to take stations outside the guard rail. The winch man misunderstood this motion and thought the mate was signaling him to let go. He lowered the pole with a run. One of the men on the lighter saved himself by diving overboard. The other lad, "Whitey," attempted to duck across the lighter under the pole. His hand was caught and badly crushed between the pole and a stanchion. "Whitey" gamely helped his shipmate to climb back aboard the lighter, then fainted.

We lowered a cargo net and hoisted both men on deck. At first it appeared impossible to save "Whitey's" hand, but the doctor gave him an anæsthetic and worked on it for three hours. Then he reported that there was an even chance it would pull through.

This incident affected the mate deeply. For an hour he was too shaken to go on with the job. I assured the mate that the accident was not his fault and suggested that he lower a small boat with two men aboard to lie well astern of the lighter on a line. After a pole was actually landed these men could haul the boat up to the lighter and swing the pole into position with a peavy. This scheme worked very well.

On another occasion the launch was approaching the passage through the reef with a heavily loaded lighter when the towline parted. The lighter immediately began to drift down toward "The Hook," the most dangerous part of the reef. If the lighter drifted into the breakers at "The Hook," it meant the loss of valuable supplies and probably death for all on board.

We swung the launch around and ran up close to the lighter to pass them a spare towline. The inexperienced lighter crew were slow getting the new towline secured. Meanwhile they drifted steadily toward "The Hook." In the excitement Barney, our launch operator, accidentally shut off the motor without realizing he had done so. When he again attempted to start it, it balked and remained dead. He went ahead on the port motor and took a strain on the towline, but with the lighter in tow the launch refused to steer. "The Hook" was very near. In mingled fear and desperation I yanked up the floor board and quickly found that the starboard-motor throttle control was loose. When we got the motor going and the tow underway again the lighter was being spattered with spray from the breakers thundering over "The Hook."

A favorite sport aboard ship, while anchored at Midway, was fishing for shark. The ship's crew put in most of their spare time at it. They made some heavy steel hooks and baited them with fish heads, attached a stout line, and watched for an unusually big shark to come nosing around the stern for refuse. Like all sailors they hated sharks and when one was hooked, a dozen willing hands would heave away and hoist the big fellow on deck. The shark is far stronger than other fish of his size and puts up a savage fight. A ten-foot shark is more than a match for half a dozen sailors until he is thoroughly tired out.

Once on deck the shark would writhe and lash out with his tail while the sailors attempted to roll him over on his back in order to slit him open. The hide of the brute is extremely tough. I have seen a sailor drive a heavy crowbar against the head of a shark without making a dent. Once a shark was ripped open the crew would shove him overboard, still struggling. The instant he struck the water he would be torn to pieces by his brothers. This sounds like cruel sport but the shark is seagoing public enemy number one. Sailors consider it their duty to destroy as many as possible.

Late one afternoon the launch returned to the ship in the middle of a shark party. The biggest shark we had yet seen was on the hook, at the stern near the boat-boom, fighting desperately. A number of other sharks nosed around looking on. It was a very rough day. The only way to get aboard the ship was over the Jacob's ladder that dangled from the end of the boat-boom. Due to the heavy roll of the ship the boat-boom pointed at the sky one moment and touched the top of a swell the next. The launch was plunging in the swells and it was an acrobatic feat to stand on her bow, make a grab for the ladder and be swung high into the air the next instant. To make the transfer safely required excellent timing and a strong grip. There were some half dozen members of our party in the launch; all hands realized that with that school of hungry sharks around us, it just wouldn't do to fall overboard. One by one they leaped for the ladder, seized it in a death grip and clambered to the *North Haven's* deck.

Many seamen believe that the term "man-eating shark" is a misnomer and that the shark is really a coward and will not attack a man in the water. That seems like bad logic to me. A hungry shark has nothing to fear as he is heavily armored and can't be hurt much even with an axe. It is true that he has a low-down appetite and adores garbage but, when really hungry, it is my opinion that he will take a bite of anything he can get his teeth into, including you or me.

There was fine weather during our first few days at Midway. We made every effort to unload our material as rapidly as possible. The beach was soon piled high with goods of every description. The lighters delivered supplies to the beach faster than the tractors and sleds could haul them to the camp site which was over half a mile away.

The tractors were equipped with Diesel engines and were enormously powerful. They would hook onto a big sled loaded with five tons of material and drag it across fifteen hundred feet of soft-sand beach, up over the sand dunes where the runners of the sled would completely bury themselves in the deep sand, without stalling.

As the material began to pile up on the beach, I became a bit uneasy. If a storm caught us we stood to lose much valuable goods. While at lunch in camp, which was sheltered from the beach by the row of sand dunes, George Kuhn came running up to tell me that a heavy squall was bearing down from the northeast. I ordered every man in camp to the beach on the double. We got there just as the squall struck. The lagoon soon began to build up a strong chop under the influence of a forty-mile gale.

Our materials were piled above high-water mark, but, driven by the wind, the water crept nearer and nearer until it lapped at the edge of the pile. We worked desperately to move the stuff to safety before the next high tide came in and swept everything away. In the midst of this a launch arrived from the ship with a full lighter-load which had to be taken care of immediately as it included perishable provisions. The wind was directly on shore, causing small breakers to run on the beach. It was almost certain that the lighter would broach when it was beached and that the breakers would then ruin the provisions. We signaled the launch to stand out from the beach and then ease down slowly, allowing the wind to drive the lighter ashore stern first. As the lighter neared the beach two men dashed through the surf and secured heavy lines to its towing bitts.   These lines were secured to the tractor and the lighter was at once hauled half-way out of the water. Other men quickly rigged heavy tarpaulins to protect the provisions from the surf.

For four hours all hands worked like mad.   Before high tide we had managed to move the last of our materials to safety.   By this time it was dark, raining hard and blowing great guns.   All of us were dead on our feet.   There was no chance to get back to the ship, but the Chinese boys had a hot dinner waiting for us at camp.   We did it justice.

The crews on the launches had not been taken off and had no way to get ashore.   They had bunks and emergency rations but I knew they were cold and wet.   George Kuhn volunteered to row out to them with hot food.   It was blowing forty miles an hour and there was a strong surf on the beach.   But Kuhn is at home in a dory and thought little of his round trip.

For two days it rained and blew and was quite cold. The tents leaked, bunks and bedding were soaked, and everyone was miserable.   The Chinese boys had a hard time keeping their fire going.   The men grumbled, but worked right on through the rain.   We sent thermos jugs of hot coffee around to the various crews.   Later I borrowed some rum from the cable company and sent around a hot rum punch.   After two mugs of it, the men stopped grumbling and started singing.

CHAPTER NINE

MIDWAY ISLANDS are designated by the United States Government as a bird refuge. We had received special instructions from the Navy Department to avoid injuring or molesting the birds. I have never seen any other place where one becomes bird-conscious as rapidly as one does on Midway. Thousands of birds circle over the islands day and night. There are birds on the ground and under the ground, and they literally get into your hair.

The beaches and the camp sites were densely populated with large birds the size of a goose, known as gooney birds. The gooneys live at Midway the year 'round except for a six weeks' migratory period of wandering. When nesting time comes to Midway each couple picks out a small hollow in the sand under the magnolia brush and the female deposits therein a large egg. The parents sit on the egg at intervals when not out fishing and with the aid of the warm sun the egg finally hatches. The young gooney that emerges grows with astonishing rapidity and within a few days is half as large as his mother. After the youngster is hatched the parents spend most of their time with him. When one of them returns from a fishing trip he expects to find the youngster in the exact spot where he left him. If he has moved ten feet away he haughtily refuses to recognize him. The young gooney realizes full well that if he changes station for any reason he will starve to death and as a result, though the heavens may fall, he stays put. This remarkable idiosyncrasy caused us much trouble and loss of time when we started clearing the station site. The young gooneys would not budge even for a tractor and we had to send a guard ahead to pick up the youngsters and put them to one side. The minute the youngster was released he would scuttle back to his stamping ground and the next tractor crew was forced to repeat the process.

There appeared to be two distinct sorts of gooneys — one a brownish black, while the other had gray back and wings, white head and breast. The black ones hovered around the ship and fought for garbage, but the gray ones had better sense and would eat nothing but fish.

The parent gooneys often collect in groups on the sand and go through a queer ritual that has all the earmarks of a tribal dance. The dance begins with the birds facing each other. First there is an energetic nodding of heads and bending of necks; after which they extend their necks, touch beaks and each bird claps his jaws together rapidly, setting up a terrific clatter, meanwhile doing an awkward shuffle.

Lastly they raise their beaks to the sky and hiss. This performance goes on endlessly all over the place. Perhaps they were making love, but as far as I was able to observe, the show was entirely platonic.

The bird that irritated us most was the "moaning bird." I don't know his real name but that very fitting cognomen was bestowed upon him by the men of the cable company. During the day the moaner lives in a hole in the ground. He flies only at night. We were forever breaking through up to our knees into moaning-bird holes and many a sprained ankle resulted.

When we turned on the camp lights at night, the moaners were attracted by the bright rays and flew right into them. We had to put screens around the mess tent to keep them out of the soup, and in the cook tent our Chinese boys were batting at them with skillets.

At night this bird lifts his voice in a moaning sound exactly like that of a human being in great pain. The effect of a chorus was a bit gruesome until we got used to it. One night I was invited over to Mr. Perry's house for dinner and on the way back I carried a lantern to light the trail. The lantern was nearly knocked out of my hand several times by moaners flying into it, and I had to hold one arm in front of my face to shield it as the birds swished by me in the night. All the while their eerie moaning racket was in full blast. I fervently wished that every moaning bird on the island would get as sick as he sounded and go ahead and die.

The moaning bird is about the size of a pigeon and is gray with a white breast. He has no especial equipment to fit him for excavation work, but can dig rapidly with his beak and claws in soft ground. It was always a mystery to us why their holes did not cave in and bury them alive as the soil was composed entirely of loose sand.

The loveliest bird on Midway is the white tern, known as the "love bird." These birds are snow white with jet black bill and eyes. They have the wing spread of a pigeon but a very slender and graceful body. They are so named because they are always found in pairs, male and female, and never separate, even momentarily. It is easy to identify a bereaved widow or widower. They sit around in a dazed fashion and soon follow their mate. These birds were chock-full of curiosity and until they grew accustomed to our presence on the island, numbers of them were forever hovering in pairs just over our heads, closely watching all operations.

The mother love bird lays her one egg in the crotch of a tree with no semblance of a nest to protect it. This is a precarious situation as any strong breeze rolls the egg off and it falls to the ground. We inspected several young love birds. They step out of the shell with a marvelous sense of balance, enter the world sitting on a tight rope, so to speak, and hold everything until they are ready to fly. I have rarely seen a bird so graceful as the love bird, and I found it hard to reconcile his lovely appearance with the fact that he lives on fish and knows nothing of music. He shows a lot of common sense at that—he can't sing, admits it and doesn't try.

The sooty tern is present on Midway in huge numbers. His shrill screeching never stopped while we were there.

He is perennially on the wing except at nesting time. I thought every bird in the world had to stop occasionally for rest and sleep but during the three weeks the ship remained at Midway no one saw a tern except in flight. Later, when the nesting season arrived, they all landed at once and proceeded to lay their eggs on the sand. They lined up on the sand shoulder to shoulder like a body of troops, faced directly into the wind and let go.

The sooty tern has a bit bigger wing-spread than the pigeon but is not so heavy in the body. They lay an egg half the size of a hen egg. We ate a few. They taste much like a hen egg when very fresh, but they go bad quickly and will not keep even in cold storage.

There were also quite a few tropic birds at Midway. Like the sooty tern, they were always on the wing. The "bo'sn bird," as he is called, is about the size of a seagull and has a forked tail. Two long thin red feathers in the tail serve to give this bird a remarkable ability to balance himself in the air. Mr. Perry insisted that the bo'sn bird could fly backward. I watched several closely to verify this. The bird heads into the wind and cuts down his flying speed to less than the speed of the wind. Balancing himself with the aid of his long tail feathers he lets the wind drift him backward. He can accomplish this trick in a fairly light breeze and actually appears to be flying backward.

Late one night our old Kanaka, Dave Richards, decided to play a joke on the boys. He stole out of his tent, gathered up a half dozen young gooneys and placed one on the chest of each of our sleeping braves. The young gooneys didn't know how to get back home in the dark so they cuddled down to wait for the dawn. When *reveille* blew the next morning each of the gentlemen who had a young gooney for a bed-fellow was ragged unmercifully.

"Well, well, can't leave the women alone, eh?"

"What's her name?"

"Just wait till her mother finds out."

Dave was finally caught red-handed and given a Kangaroo Court.

Mr. Perry told us there had formerly been several donkeys on the island, but they continually broke into the vegetable garden and ate up the flowers in the compound. They were of little use and were transferred to Eastern Island where they reverted to the primitive and became wild.

Eastern Island sounded interesting so the following Sunday several of us explored it. We failed to sight the donkeys because of the thick brush but we did find some deep holes on the beach which they had pawed in order to find brackish drinking water. The gooneys were thick on Eastern Island. The beach was so packed with gooney families that we had to pick our way carefully to avoid stepping on the youngsters. On the far side of the island we found a number of hollow-glass balls of assorted sizes. These had floated across the Pacific from Japan, and made very acceptable souvenirs. The Japanese use these glass balls in lieu of cork floats for their fish nets, as cork is expensive in Japan while glass is cheap. These balls range in size from three to fourteen inches in diameter and can stand a lot of punishment. They had had to cross the reef through heavy surf to reach the island.

That Sunday, after dark, we gave Midway Island its first movie show. The only man who failed to attend was the cable company's operator on watch. We had brought two movie and sound machines along, one each for Midway and Wake, and a number of programs. The feature picture for the evening was "Red Hot Mamma." Judged by the comments from the crowd it was a box-office hit. It certainly lived up to its title.

Mr. Lawn, the cable-company engineer, had not seen a movie for six years, and it was rumored that he preferred life at Midway because of the complete absence of feminine society. But for several days after seeing "Red Hot Mamma," his high resolve appeared to be slipping, the evidence being a thoroughly abstracted manner, frequent long sighs and an occasional vague smile. Mr. Perry said if we showed another movie like that one, his Chinese boys would head for Honolulu on the next ship.

CHAPTER TEN

ON the morning of April 16 we received a radio message from Alameda which read as follows: "The *Pan American Clipper* will depart Alameda at 3:50 P.M. today for Honolulu stop crew Musick, Sullivan, Noonan, Canaday, Wright, Jarboe." For the rest of that day and night the radio office was the center of interest at Midway.

We were four hours west of Alameda, so it was 11:50 A.M. at Midway when Musick took off. At lunch time the construction crew got the news that the *Pan American Clipper* was flying to Honolulu. They had never seen the *Clipper* but you would have thought each of them owned a share in it. After lunch they all trooped over to the radio office to hear the latest news. They went back to work reluctantly when I promised to send a messenger around with the radio bulletins as fast as they came in.

The Midway operator tuned in as soon as the *Clipper* took off from Alameda and copied every message she sent or received. These messages were immediately posted on our bulletin board. Every half hour Musick sent a position report to Alameda which included his altitude, course and a short weather report.

Shortly after leaving Alameda he told them he was flying at ten thousand feet with blue sky overhead and a white cloud bank below. His radio operator called Alameda and said, "Stand by to give us a bearing." Then he pressed down his key and sent a continuous signal for one minute while Alameda's direction-finder station tuned on him. Then we heard Alameda answer, "You bear from us two hundred fifty-two degrees."

All that night we listened in from Midway, closely following every move the *Clipper* made, silently "wishing" her along. We heard her talking to several ships along the course. Once we heard her say to a ship, "Can see your lights through a break in the clouds. What is your position?" The ship answered with her position and the *Clipper* said, "Thanks." Musick told Alameda that he was averaging one hundred and thirty-six miles per hour, which meant that he was flying well throttled down to save gasoline.

Halfway across, Honolulu took over the job of giving the plane its radio bearings. We heard Honolulu tell the *Clipper* that Army and Navy planes were coming out to meet her and escort her in. Early the next morning we heard the *Clipper* tell Honolulu that the aerial escort had found her and joined up. Then we heard her say she was circling over Honolulu. She was scheduled to arrive at Honolulu at 8:00 A.M. Five minutes before eight we got the landing signal from her operator. We were happy that the first big jump had gone off so smoothly.

Later that day the Honolulu operator told us that the arrival of the *Clipper* was being celebrated all over town, and warned us to stand by for Musick's broadcast. We tuned in on the broadcast and heard Musick say that the flight was made possible by the joint efforts of the entire aviation industry, and offer tribute to the brave pilots of earlier days who had been lost at sea *en route* to Honolulu.

We went back to our work.

After a week's visit the *Pan American Clipper* took off on the return trip to Alameda. Again we copied her radio messages and held thumbs for her as she bucked headwinds across the Pacific, arriving some minutes late.

Alameda radioed us a day later that the next flight of the Clipper would be from Alameda to Midway as soon as the latter station was ready. This news caused us to re-double our efforts to speed up the construction program. By the end of our second week at Midway it was well along.

Mittag and half of the construction crew were to remain at Midway and complete the job, while the *North Haven* was to depart for Wake Island as soon as all Midway sup-plies were ashore. But the perishable provisions could not be unloaded until the Midway cold-storage plant was in commission. So we concentrated on that detail.

The plant required a dependable supply of cooling water for the compressors. This entailed the completion of a well, five feet square by fourteen feet deep and cased with redwood. It furnished an ample supply of brackish water, which was satisfactory for all purposes except cooking and drinking. For the fresh-water supply we must depend on rain water, which would be led from the roofs of buildings to underground cisterns.

Water pressure would be supplied from the two windmill towers, sixty-five feet high. Each tower had an eight-thousand-gallon redwood tank on top, one being for fresh water and the other for brackish. The towers were set on heavy, six-foot concrete foundations.

The radio station must also be completed before we departed. This installation included long underground cables, telephones and a number of fifty-foot antenna poles which had to be accurately set and aligned.

The work of unloading the ship was pushed ahead as rapidly as possible. When the sea was too rough to permit unloading heavy material, we handled gas and oil for the planes. The long busy days passed quickly. From daybreak to dark we stopped only for meals. As soon as the cold-storage plant was in commission we sent ashore the provisions, and shortly thereafter completed the discharge of all Midway cargo.

Before the *Clipper* arrived, Midway must be completely ready. "Ready" meant that the seaplane dock, landing float, refueling facilities, buoys and anchors for handling the *Clipper* must be complete. The landing and take off areas in the lagoon had to be carefully surveyed and sounded to insure that the *Clipper* would not run afoul of a hidden reef or coral head. Furthermore, Midway must be prepared to furnish bed and board to the flight crew, and the radio direction-finder must be ready to guide the *Clipper* to Midway. The *North Haven* still had her biggest job to do on Wake Island. The final steps at Midway were up to the engineer and crew who were to remain there.

After landing the Midway cargo, it was necessary to return to the *North Haven* the Wake tractor, launch, lighters and camp gear which had been pressed into service to speed up the work. The sea was rough and we had a difficult and dangerous time getting this heavy gear aboard. On one occasion a lighter lying alongside the ship snapped its head lines while surging in the swells and started to drift astern with the stern lines still fast. The ship was turning over her propeller slowly to hold her head up into the seas. It could be seen at a glance that when the lighter took up the slack of her stern lines she would be swung under the ship's stern and into the revolving propeller. This would seriously cripple the ship. There was not time left to throw off the lines, but the mate grabbed an axe and cut both of them at the ship's rail. The lighter drifted astern and missed the propeller by four feet, as the lighter crew prepared to dive overboard. Since the Wake tractor was on that lighter we would have been forced to return to Alameda for another one had it been lost. The launch overhauled the lighter and towed it alongside again.

At daybreak on April 30 we had completed loading the *North Haven* except for the two launches and the Wake crew. We started out from shore soon after daybreak in the big launch and headed for the northwest passage.

The sea had built up during the night. When we neared

the passage we saw that we could not get through as it was blocked by a foaming white line extending from "The Hook" to the North Breaker. We eased up as close as we dared and watched for a few minutes as the huge swells reared their heads at the reef, curled forward and crashed down in a wild smother of spray and foam. There was no telling how long this condition would last. We were already three days behind schedule.

We headed for the beach. On the way in I decided to make a survey of the southeast reef and endeavor to find a safe passage for the big launch on that side, which was the lee. Arriving at the station we took the light-draft launch and set out to feel our way through the southeast reef. After several hours' work, we found a passage that was possible but hazardous. The big launch would have to lay to, watch for a calm period and then dart across. If a swell ran out from under her as she crossed the reef, she would snag her bottom. We marked this passage with can buoys.

The *North Haven* was directed by radio to anchor off the southeast passage. As soon as she was in position, the light launch made her way safely through the newly marked channel. I then returned in her to the station and sent out the big launch with the construction crew and a lighter-load of camp gear for Wake. The light launch was used to transfer the crew to the ship and then the big launch started out with the lighter in tow.

At the reef we hauled the lighter up close astern of the

launch and coiled up the slack of the towline free for running. At a favorable moment the launch rushed through the passage under full power and payed out the towline rapidly until she was across the reef, where she slowed down and took up the slack gently. As the lighter only drew about sixteen inches of water, there was no danger of her striking bottom.

With the camp gear aboard we were ready to sail for Wake Island as soon as the radio direction-finder at Midway was calibrated. This was begun early next day.

The radio direction-finder is a device which locates the exact direction from which a radio signal is coming. As a ship or a plane approaches Midway, the radio operator presses down his key, thus sending a continuous radio signal. The radio operator at Midway turns his tuning dial until he picks up the incoming signal accurately. He reads off the bearing or direction from his dial and sends this information to the ship or the plane, so that it may know the exact course to steer for Midway. There frequently are errors present in direction-finders. The process of determining these errors is called calibration.

The *North Haven* headed out to a point about five miles from the station and slowly steamed in a circle, keeping the station in the center. At intervals she sent a radio signal which the Midway operator would tune in and mark on the dial. Simultaneously one of our engineers ashore took the exact bearing of the ship through his surveyor's transit, from a tower just over the radio station. The engineer's results were accurate. When compared with the bearings obtained by the direction-finder, the errors were easily obtained and would be allowed for on all future bearings.

Calibration required over half a day. The Midway launch came out bringing the rest of our party and several of the cable company's staff who had come to say goodbye. After hearty "so longs" all around, the launch pushed off. We got under way for Wake Island at 2 P.M. on May 1.

Behind us lay our first small colony — isolated but independent. They would be quite comfortable in a few weeks when the quarters were completed. They would have electric light and power, hot and cold running water, an ample cold-storage supply of fresh provisions, a library, and movies twice a week. Soon they would be ready to receive and service the *Pan American Clipper*. Within a year's time hundreds of people would pass through Midway.

CHAPTER ELEVEN

WAKE ISLAND was five days away. Everyone on the *North Haven* was glad of the chance to relax. There was much good-natured chaffing and swapping of yarns, and many tall stories were floating around. The boys tried to palm most of the latter off on Junius Wood, but he was suspicious and refused to write about anything he had not actually witnessed. Dan Vucetich and his gang had been too busy unloading the ship at Midway to get ashore. The shore crews told them the story. All hands were in excellent physical shape except Bill Young, who had picked up a bad cold during a rainy spell ashore. The cold had developed into pneumonia.

En route to Wake our secret brotherhood, the "Ancient and Royal Order of Gooneys," held several meetings. One night shortly before we left Midway I had been invited to join a select group who were to meet in the newly completed mess hall. The avowed purpose of this gathering was to give our far-famed A. G. A. stoves a speed run, and see if they would turn out a good steak. Some ten or twelve men were present when I arrived, and I also noted the presence of a pair of large gooneys who were waddling about the hall and inspecting everything with much curiosity. After a delicious snack of broiled steak, potatoes and coffee, Bob Kratz proposed that we found forthwith, the "Ancient and Royal Order of Gooneys." Officers were elected, by-laws drawn up, and those present were declared charter members. Additional members were to be solicited as soon as time permitted. Initiations would be held on the ship when we left Midway.

The third day out from Midway all candidates were invited aft to undergo the trials and ordeals preparatory to entering this noble lodge. They were blindfolded and put through ingenious forms of torture at which Dr. Kenler, in his surgeon's smock, presided as high priest.

Our boat-boom was still in place. As a crowning touch all candidates were made to walk the plank blindfolded, each man thinking that he was out over the ship's side on the boat-boom. To simulate this hair-raising effect they were herded into the after galley and a facsimile of the boat-boom was rigged flat on the poop deck. The candidate, securely blindfolded, was brought up from the galley, and after being violently spun around, was assisted to climb over a false ship's rail onto the false boat-boom. He was then given a strong push. As the victim let out a yell of terror and fell to the deck, he received the contents of a large bucket of water in his face.

The *North Haven* plowed along on her course under fair skies. Our crew was kept busy overhauling equipment and making up new fenders and towlines for the lighters. In the wardroom mess we completed final plans for the battle of Wake Island.

Bill Young was growing steadily worse. Dr. Kenler reported that he was dangerously ill and would probably die unless we could get him to a hospital. I immediately radioed this information to Alameda and requested instructions.

The *North Haven* was at that time equidistant from Honolulu and Guam. I knew there was a Naval hospital at Guam and that Wake Island was much closer to Guam than to Honolulu. Guam looked like our best bet, but I was not sure that we could get more fuel oil there. The ship was chartered for a definite period. We had to complete the job within a time limit, and every day was precious. But a man's life was at stake.

Just then, like a scene from the movies, the United States Navy stepped into the picture and came to the rescue. The Naval Transport *Henderson* was *en route* from Honolulu to Guam, and within one day of the latter port, when the Navy Department ordered her to turn back and meet us, pick up our sick man and take him to Guam. Alameda had been busy. They received my radio at 2 P.M. At 10 P.M. the captain of the Henderson radioed the *North Haven,* giving his position and speed, and requesting us to meet him on the Great Circle Course from Wake Island to Guam. We put on our last ounce of speed and started in to do some close navigating. Two ships rushing to meet each other in the middle of the Pacific Ocean — tiny specks on a limitless waste. If either ship made the slightest error in navigation they would fail to meet.

I checked our course and position half a dozen times a day. From time to time we radioed our position to the *Henderson* and she answered with hers. Both ships held dead on the Great Circle Course. When we finally picked up the *Henderson*'s lights she was so nearly dead ahead that the two ships would have collided if they had held on.

We met about 9 P.M. and our crew had a chance to witness some smart Naval seamanship. The *Henderson* put her doctor and stretcher-bearers into a thirty-foot motor launch, with all hands in lifejackets. The boat was lowered and took the water easily. We had lowered our port gangway, and the Navy launch showed nice judgment in coming alongside in a fairly rough sea. Their doctor came aboard with his stretcher-bearers and the launch hauled out and lay to.

Dr. Kenler took the Navy doctor in to see Bill Young and gave him the history of the case. Bill was apparently a little better. He was pleased and excited when we bundled him up warmly and strapped him into the Navy stretcher.

Meanwhile our captain maneuvered his ship into a position that would give the best lee possible for the Navy launch.

We eased across the bow of the *Henderson* so close aboard that you could almost have jumped from one ship to the other.

The Navy launch came alongside again and two husky Navy men carried the stretcher down the gangway and slid it into the launch without the least jar.

The Naval doctor said goodbye and jumped aboard the launch which put back to the *Henderson's* side, under the big crane where they hooked on and were hoisted in.

We gave the *Henderson* a whistle salute which she returned, and the ships got under way. A heavy rain squall blotted them from view as the *Henderson* headed for Guam, the *North Haven* for lonely, uninhabited Wake Island, the only spot of land within a radius of twelve hundred miles.

CHAPTER TWELVE

Our charts showed Wake Island to be shaped like a horse-shoe, with a lagoon in the center. It is entirely surrounded by a wide reef. The waters around the island are so deep that a ship cannot anchor. There is no channel through the reef, though one spot on the chart is labeled "the boat landing." I had been informed by Navy personnel that two Navy boats had been wrecked in the surf at that point.

I organized two survey parties to investigate Wake; one to make a land survey, the other a marine survey. The shore party was ordered to take emergency rations with them in case they were cut off from the ship by bad weather. On May 9 at 11 A.M. we sighted Wake Island, a dim smudge in the distance.

The *North Haven* stood up in the lee of the island, which was the south side, and laid to off the spot marked on the chart as the boat landing. Climbing to the crow's nest with my glasses, I saw that there were really three islands. Each leg of the horseshoe is cut through by a shallow channel. These three islands are named Wilkes Island, Wake Island and Peale Island. The boat landing was on Wilkes Island which lay opposite us. A small channel (Wilkes Channel) runs through the island adjacent to the landing.

Wake Island forms the upper part of the horseshoe and is by far the largest of the three. Across the lagoon from Wilkes Island is Peale Island. It is cut off from Wake Island by a small channel known as Peale Channel. The islands appeared to be covered with scrub trees, brush and jumbled masses of rocks. We put overboard the light launch with the surf boat in tow, and sent our survey parties ashore. Russel had charge of the shore party, while the first mate and I were to make the marine survey.

There was an excellent lee at the time and very little swell, a welcome change from Midway. The boat landing was calm.

We landed our shore party without difficulty and noted that the beach was indeed a mass of rocks and boulders. The water was crystal clear and alive with brilliant colored fish. We rubbed our eyes and stared at this display —fish of vivid green, orange and black, pale yellow, and by all the gods, Mr. Ripley, one with distinct red, white and blue stripes on its sides!

The shore party were equipped with emergency rations, water and machetes, and at once struck off into the brush to explore Wilkes Island, while the mate and I started to cruise around the group with the surf boat in tow. We inspected the entire shore line without finding a break in the wide solid reef. The mate took hundreds of soundings in the endeavor to find a spot where the ship could anchor, but there was no bottom. Apparently these islands are the top of a marine mountain.

The reef runs straight out from shore for several hundred yards on a level with the beach, then slopes sharply down for perhaps two hundred feet, and then ends abruptly.

The water was so clear we could plainly see where the reef stopped at a depth of eighty feet.

The north side of the islands was the weather side. We did not dare get in too close to the reef there as the breakers thundered against it and we might be sucked in. A few birds flitted to and fro. There were no other signs of life.

After circling the islands we returned to the boat landing, convinced that it was the only possible spot where we could land supplies. It looked far from favorable.

The beach was a tumbled mass of coral rock, six feet or so above sea level. About a hundred feet back from the water's edge the jungle started; queer looking thorny stunted trees and thick brush. Clearing ground on Wilkes Island was going to be a laborious process. Transporting the gear from the landing to the station site would not be easy.

It was evident that as long as the boat landing remained the lee side we could land goods, but the least bit of weather from the south would stop us completely.

I walked up the beach a couple of hundred yards and inspected Wilkes Channel. This channel led from the lagoon across the reef to the sea. It was very shallow, the depth of the water varying from six inches to two feet.

The shore survey party returned about this time. As night was coming on we all returned to the ship and hoisted in our boats. The ship stood off a couple of miles from the reef to have a safe amount of sea room. We had hung a lighted lantern in a tree on the beach so the ship could keep the island spotted during the night.

The shore survey party had a dismal story to tell. They were well scratched up by the thick brush and reported that they had been all over Wilkes Island. It was no place to build an air base because it showed plain evidence of having been under water. They had found logs in the center of the island that could only have been washed up by the sea. In many places driftwood was lodged in the trees overhead. The entire island was covered with jagged rocks and boulders and thorny brush that would render the construction of a station difficult. And finally there was shallow water on the lagoon side of the island for a long distance out from shore ; and this water was so full of coral heads that it would be very difficult to get out through them by boat to service a plane.

I listened to their report and pointed out that we had to have a station here even if we were forced to build it on piles to protect it from high water. We would explore the other two islands and pick the most favorable location. But whatever the difficulties were, the station must be built, as the transpacific airline could not operate without stopping at Wake for fuel and service. We decided to build a dock at the boat landing as the rocks would ruin a lighter if beached there.

Orders were given to start the dock construction at daylight. The crew worked late on the ship, getting the materials ready for the job. George Kuhn was placed in charge. He and McKenzie worked out a plan for a dock that proved to be very satisfactory. The next morning at daybreak the ship moved in close to the boat landing and we put over a lighter and the light launch. We put down heavy kedge anchors, weighing eight hundred pounds, for the lighters and launches, with heavy buoyed chains so that there would be no danger of our craft dragging anchor even in stormy weather. Then we sent over the materials for the dock, and work was begun at once.

McKenzie was placed in charge of a party to explore Wake Island, while Russel and I elected to investigate Peale Island. In order to get to Peale we sent a skiff from the ship to the boat landing, carried it overland to Wilkes Channel, and then rowed across the lagoon to the island, a mile and a half away.

We landed on a sloping sand beach, hauled the skiff out of the water and started across the island. It was at once apparent that Peale Island was a different place altogether from Wilkes Island. There were no rocks and the soil was a rich brown loam. There were fair sized trees and many vines, and no evidence whatever that the island had been under water.

I had no idea that there were moaning birds there until I broke through to my knee in one of their holes. I had thoroughly disliked the moaning birds at Midway, but now I could have wept for joy at finding them. I knew full well that those birds would never dig holes where they might be drowned out by high water. Russel and I did a war dance and whooped with delight. We ran about looking for more holes to be sure that the first moaner was not all by himself, and perhaps misguided. We found plenty more of them and had no further worries about high water.

We crossed over to the ocean beach and looked down on a lovely view. There was practically no beach as the ground terminated abruptly in a huge natural breakwater of ancient weather-beaten coral rock twenty feet high.

This breakwater sloped sharply down to meet the reddish coral reef. The reef stretched out into the sea for several hundred yards and then sloped out of sight. The breakwater gave the appearance of having been man-made as it was very uniform in height, slope and composition. Yet we knew that every rock had been driven into place by the restless action of the sea throughout the centuries. Perhaps "forty centuries looked down" but here was no man-made thing, and nature had furnished no ruler by which to measure the time it took her to build this mile-long monument. With that wide reef to crumple the big seas and the breakwater to fend off any surge across the reef, we could see why the moaning birds felt confidence in Peale Island as a place to live.

On the upper end of the island we found the remains of a Japanese shack that had probably been the home of a lone fisherman. Russel poked among the débris and hauled out a queer-looking old water jug which he regarded enthusiastically. He acclaimed it an historic relic and decided to have it classified by competent authority at the first opportunity.

Well satisfied with our discoveries we rowed back to Wilkes Island and hauled our skiff out on the lagoon beach. George Kuhn had made excellent progress on the dock and said it would be ready for use by the following noon.

McKenzie returned with his party thoroughly discouraged and scratched up but when they heard our report on Peale Island everyone's spirits lifted.

That night on the ship we sat down to consider the job ahead. We were reasonably sure we could land our material on Wilkes Island, but in order to ferry it to Peale Island, we had to get a lighter and a launch into the lagoon. It seemed that the only way to accomplish this was to build a skidway across Wilkes, haul the lighter and launch out at the boat landing, and skid them across inch by inch.

The rugged nature of Wilkes Island would render such an undertaking a long and difficult job. Time was short. However, there seemed to be no other way.

SKYWAY TO ASIA

We decided to clear a temporary site on Wilkes Island and set up camp immediately. Then we would clear a site for a storage yard in the center of Wilkes and gather all materials there. Meanwhile Russel would take a survey party over to Peale and lay out the site for the station and mark the place where each building was to go. I would take charge of getting all material ashore and figure out some way to get a launch and lighter into the lagoon.

At daylight of the third morning we started to clear brush and pitch camp. Bicknell, an ex-Army officer, set up our Army tents and laid out a company street. Axes were soon ringing, saws humming and dirt flying. George Kuhn had his dock ready on time. It was a sturdy affair set up on shear legs constructed from our spare redwood radio poles. The stringers were also radio poles. Over these were small railroad rails, covered in turn with two-by-ten planks. That dock was fifty feet long and was built in a day and a half. It would have supported a locomotive.

At the outer end of the dock was a landing stage consisting of two small scows decked over, lashed together and securely anchored at each corner. The first lighter with the tractor aboard arrived at low tide when the top of the dock was several feet higher than the lighter. Caterpillar tractors are familiarly known as "cats" and a tractor driver is a "cat-skinner." Gill was our cat-skinner. He nursed that tractor up an improvised gang-plank at an angle of forty-five degrees, and onto the dock, while the landing stage surged back and forth and threatened every instant to topple the tractor overboard.

The next thing sent ashore was the dog house — a small radio shack, completely equipped for immediate service. The tractor hauled the dog house to the camp and in half an hour we were getting messages through to Alameda via Midway and Honolulu. In two days' time the camp had tents set up and floored, including a mess tent with tables, a cook shack, commissary tent, temporary power house and latrines.

The ship loaned us a big kitchen range for temporary use, and a hot water stove and boiler. We sent ashore a supply of provisions, a small electric refrigerator and cots and bedding, and thereafter the camp was independent of the ship except for fresh water.

We had started digging a well the first working day but struck rock immediately and were forced to blast our way through the hard coral. The well was located just behind the kitchen. We had to put in small charges and blanket the hole to avoid injuries from flying rock when each charge was detonated.

It was hot as an air-cooled cylinder-head and the camp was screened from any breeze by the surrounding brush. Some of our crew were knocked out temporarily by the heat. When we finally got the well down we had a boundless supply of pure sea water.

The porous rock strata allowed the sea water to seep through. The water in the well actually rose and fell with the tide. We gave the well up in disgust and ran a pipe line to the lagoon to furnish salt water in camp for bathing.

CHAPTER THIRTEEN

THE cable company at Midway had kindly provided us with a dozen hens and a rooster with which to found the first Wake Island chicken colony. Dave Richards had chaperoned the chickens on the trip across and remained in charge of them at Wake. We built a chicken-run ashore, and Dave took his charges over to the island and set them up in business.

In addition to this responsibility, when we started to blast the well, Dave had charge of the "powder." His only failing was that he was too generous with his charges. His idea was, "one big charge and there's your well." I finally persuaded him that this would wreck the entire camp and he went ahead more cautiously.

Due to the failure of the well, we had to face an acute fresh-water problem. The needs of the camp and the survey party on Peale Island must be supplied from the ship. A special work party was required to distribute fresh water to the various gangs. This was such a slow and laborious process that the men drank up the water faster than it was supplied. We experimented with the electric stills we had brought along, but they turned out to be so temperamental they were useless. I decided to construct a still.

My first idea was to utilize the exhaust heat from the Diesel electric generator to boil the sea water. I went ashore to look the plant over. I found to my surprise that the exhaust from a Diesel motor is quite cool. This provided the gang with a good laugh at my expense, and caused me to seek expert counsel. George Robinson was a master plumber and must know all about stills. I got George in my room and we went into a huddle:

"George, did you ever build a still?"

"Yes, sir."

"What kind of a still?"

"Well, sir, it would make either whiskey or alky or brandy but I guess they mostly made alky."

"H'mm — could you build a still to make fresh water out of salt water?"

"Sure I can."

"What are we going to use to fire the still with?"

"Dry brush is plenty good and old Dave Richards ought to be a good fireman."

So George steamed out some empty gas drums and built a fresh water still that Dave Richards fired successfully. It turned out a hundred and fifty gallons of fresh water per day — enough to make the camp independent of the ship as far as fresh water was concerned.

As soon as the storage yard was completed we concentrated on unloading the ship. We made much better time than at Midway because there was a shorter distance to tow, and the sea was much calmer.

Captain Borklund was thoroughly annoyed, however, by the fact that the ship was forever drifting away from the island or steaming back up toward the boat-landing. He took numerous soundings but apparently even then was not satisfied that there was no bottom, so one day he squared away and let go his anchor just to make sure.

The mate stood by on the fo'c's'le until the last fathom of chain ran out. Then he stopped the anchor engine. The captain tore his hair and swore steadily in Swedish while the ship drifted serenely on. As soon as the captain cooled down I asked him if he thought it would be safe to ease the bow of the ship in over the edge of the reef and drop anchor on the reef, trusting to an off-shore breeze to keep us clear. The captain, fearful of putting his entire ship on the beach instead of only the anchor, was by no means keen on this idea. But finally, weary of drifting, he tried it. The ship eased in toward the island with the mate on the bow staring down into the depths, watching the edge of the reef slide under him. He yelled "let go" and down went the anchor on the reef. The mate let out a lot of chain and the ship drifted back from the anchor and held her position nicely. At night we upped anchor and steamed off shore for safety.

As soon as this became routine, our unloading schedule speeded up. Our stern was only three or four hundred yards from the dock and with two lighters in use we kept the unloading crew busy. The tractor hauled the freight from the dock to the storage yard using the big sleds which worked as well over the rocks on Wilkes Island as they had on the sands of Midway. At the storage yard all material was stacked in an orderly fashion, window frames in one pile, door frames in another, so that when construction began at Peale, any item could be reached without confusion.

Other transportation problems now began to give concern. We only had one tractor. When we began to ship goods to Peale this tractor would be needed there. Yet it would also still be needed at the yard in Wilkes, as a large part of our materials were much too heavy to transport by hand. We could hardly cut it in two. Also, we still had no launch or lighter in the lagoon.

The storage yard at Wilkes was in the center of the island; from there to the lagoon the ground sloped gently. If we could devise tracks and some kind of car on wheels we could load the car at the yard and roll it downhill to the lagoon. The car would be easy to push back uphill when empty. Such a scheme would make it possible to release the tractor to Peale after the goods were all in the storage yard.

I remembered that we had brought along two sets of Ford automobile wheels and tires which were to be used in making up trucks to handle baggage on and off the planes, when regular transpacific flight service started. Inspecting these wheels I noted a deep wide groove in the center of the rims. This suggested that the wheels would ride a rail. We found the proper-size rail among some

iron stock brought along for use as braces, stringers or
dead men. So we proceeded to build a railroad two hundred yards long, which Junius Wood promptly dubbed the
"shortest railroad in the world."

Our little railroad was primitive but it did not operate
at high speed and was safe enough for our purpose. The
biggest job by far in building it was clearing a path through
the jungle and leveling a road-bed. Boulders as big as
a cook stove had to be moved or blasted out. Hundreds of
trees were cut down and the stumps sawed off level with
the ground. Once the road-bed was finished we laid two-
by-ten planks down lengthwise and tied them together
every ten feet or so with a two-by-ten spacer. The iron
rail we laid carefully, spaced and tacked down with
twenty-penny spikes bent over on each side every foot or
so. The rails were bolted together with home-made fish
plates and stove bolts. This construction may sound
flimsy but reflect on these facts : our railroad had no col-
lisions, no derailments, no open-switch wrecks, no casualties
whatsoever. Can any railroad elsewhere boast of such a
record ?

Once our railroad was constructed we proceeded to supply it with rolling stock. The Ford wheels were set up
in a wooden jig on the track, with the grooves in the wheel
rims fitting down over the rails. Held thus firmly in
position we brought down a portable welding outfit and
welded the axles in place. The axles were made from
heavy three-inch pipe and welded to the spindle of each

wheel. This gave us a wheel assembly with ball bearing wheels and very strong axles. It was a simple matter from here on to build and fit a strong wooden frame over the axles. The completed car was twelve feet long, six feet wide. It was equipped with a brake to keep it from running away downhill and was so light that when empty one man could push it up the slope from the lagoon to the storage yard. It carried many two-ton loads of freight.

CHAPTER FOURTEEN

THE situation on Peale Island was getting out of hand because of lack of transportation in the lagoon. We had a dozen men on Peale clearing ground, laying out building sites and digging a well. These men had to row across in the morning, taking food and water with them. Extra drinking water had to be sent to them through the day and there were numerous other errands to Peale. The round trip in a row-boat required over an hour.

Day after day we had studied Wilkes Channel and the reef at low tide, hoping to find some way of getting a launch into the lagoon. The reef was completely exposed at low tide and had only a foot and a half of water over it at high tide. But we were due for a full moon very shortly which would bring extra-high tides. After studying the matter carefully I became convinced that with good luck we could shoot a lighter across the reef at full high tide and work it through Wilkes Channel into the lagoon. The lighter only drew about sixteen inches of water. I planned to load the light launch on the deck of the lighter, strap it down securely in a cradle and send the whole business across the reef at once.

The success of this plan depended upon timing the crossing at the reef so that the lighter would be lifted by a swell at just the right time to jump over a hog-back section of the reef that was the highest and most dangerous point. Should the swell run out from under the lighter at the wrong time and drop it on the reef, it meant goodbye to the lighter and the launch, and for the present, at least, to Wake Island and the transpacific airway. But we were pressed for time, and we had become used to weighing chances and of accepting them when the odds were even.

Having decided to accept this one, a dozen men were put to work clearing a path for the lighter through Wilkes channel. Tons of small rock were picked up and thrown to one side. In many cases deeply buried rocks refused to budge and had to be blasted out. Here Dave Richards, powder king, took charge.

Dave had three men in his gang to drill holes for him and a ton of dynamite to draw from. He would drill a hole underneath a rock, big enough to hold several sticks of dynamite. Then he would pack the dynamite into the hole, tamp it down and attach a fuse cap to which were attached the ignition wires leading to the magneto. The powder gang would back off about five hundred feet, unreeling their ignition wire. When everything was set, Dave would yell "powder!" at the top of his voice and look around to see that no one was in the danger zone. Then he would take the magneto crank from his pocket and give the magneto a twist. The shattering bark of the exploding dynamite was accompanied by a shower of small stones and water. The big stone was torn loose from its bed, and rolled to one side.

Underneath the loose surface rock in Wilkes Channel was a solid bed of dense table coral as hard as granite. Dave drilled down several feet into this and put in a couple of sample shots to see how difficult it might be to blast a deep water channel all the way through the reef. It took all day to drill a hole three feet deep into that rock. Dave packed the hole full of dynamite and let her go. The hole that resulted was about a foot in diameter. We gave up that idea.

Wilkes Channel was full of gayly colored fish, eels and shell fish. The gang clearing the channel often became so interested in this marine life that they forgot the job on hand. But finally they had a path cleared through the channel twenty-five feet wide and down to bedrock.

At low tide we drove a heavy iron bar into a crevice at the outer edge of the reef. To the iron bar we attached a strong Manila line which we then led across the reef to the channel. A lighter was laid alongside the ship and the light launch was hoisted aboard it and set in a cradle. Both cradle and launch were lashed to the deck of the lighter.

At high tide every man available was stationed on the reef to help haul the lighter across when the signal was given. At the peak of the tide the big launch took the lighter in tow and led it to a point outside the reef just opposite the iron bar.

George Kuhn in the surf boat took the end of the line from the iron bar and made it fast to the stern of the lighter. The other end of this line was held by the crew inside the reef. The launch allowed the lighter to ease down to the danger point.

There was flat water for an instant between swells at the hog-back and then an extra big swell came along. George Kuhn watched it come. Just before it reached the lighter he swung his arms over his head and yelled "Take her away!" Instantly the launch cast off her line and the crew ashore hauled away for all they were worth. The lighter rode across the hog-back without a scratch. A minute later it was well inside the surf. In a few minutes with every man pushing and tugging, it started through the channel leading to the lagoon.

In the channel it rubbed bottom occasionally and finally came to rest on a shallow spot and refused to budge. Whereupon Gill ran his tractor out into the channel and hauled the lighter on into deep water.

The channel was full of holes waist-deep, and before the tractor came out we searched out these holes and stood a man in each hole to act as a temporary danger marker. After much tugging, pushing and swearing, we got the lighter into the lagoon and out into waist-deep water. Then we rigged a skid and slid the launch from the lighter into the quiet waters of the lagoon.

At last we had transportation to Peale Island. It is said that civilization advances with transportation. Peale

Island developed rapidly. There was quite a real-estate boom and lots were staked out all over the place.

Wilkes Island quickly dwindled to a second-class neighborhood and it became the ambition of all inhabitants to move to Peale.

The "shortest railroad in the world" had more business than it could handle and paid big dividends. The railroad terminated on a small dock at the lagoon beach. The freight car would come rolling down the slope with a load of light material aboard and stop on the dock where the material was transferred to the lighter. Then the launch hooked on to the lighter and towed it across to Peale.

On the beach at Peale there was a flat ledge of coral that jutted out into the water a short way and formed a natural dock, where the lighter unloaded.

Our success at getting the lighter across the reef inspired several other bright ideas. Included in the equipment list for the station at Wake were seven huge steel fresh-water tanks. Each of them had a capacity of four thousand gallons and was twenty-five feet long and six feet in diameter. To transport those tanks across Wilkes Island would be a long and difficult job, and we therefore decided to float them across the reef and through the channel into the lagoon, as when afloat they would not draw more than twelve to fourteen inches of water. However, each tank had a stand pipe and manhole in the center of one side. When the tank was placed in the water this extra weight would immediately cause the stand pipe to swing down under the tank where it would would snag on the reef. We got around this difficulty by making a saddle of two fifty-foot radio poles for each tank. When lashed between two poles, the tank with stand pipe uppermost was prevented from turning.

Each high tide we sent a couple of tanks across the reefs until they were all safe in the lagoon. We also made rafts of the radio poles and sent them through the reef and across the lagoon to Peale. Lou Patten, our number-one lumberjack, was in charge of making up the rafts alongside the ship. He handled those big poles in the water as if they were knitting needles.

The job at Wake presented many unusual problems. Each time we found some man in the crew with the necessary experience to solve it.

CHAPTER FIFTEEN

WITH one lighter in the lagoon we had only one with which to work the ship. So the unloading went slowly. The crew at the dock had much spare time in which to fish and explore the reef for specimens of shell fish and coral.

Among the dock gang was a young fellow named Mullahey who had joined us at Honolulu. Mullahey had been born and raised in the Hawaiian Islands and was more at home in the water than most natives. He spoke the language of the Islands and was fully conversant with native customs and superstitions. Mullahey's favorite sport was shooting fish with a sling and an arrow, and the dock crew were intrigued beyond words at his performance under water.

The water was crystal clear and Mullahey's every move was clearly visible. The gang would point out a particular fish they wanted brought up. Mullahey would slip on his water goggles, and slide into the water with his sling and arrow in one hand, leaving the other hand free for swimming. He would ease up to within a few feet of the chosen fish, fit his arrow to the sling and let fly. He rarely missed — time after time bringing up fish of all varieties. The crew tried their hand at the sport with small success, as they could not maintain their stance for a shot nor could they aim the weapon properly. Shooting fish was not so simple after all.

Mullahey's sling consisted of a piece of hollow cane about eight inches long to which was attached two strong rubber bands, the bands being connected at the outer end by a leather thong. The arrow was a slender steel rod four feet long and a quarter of an inch in diameter, with a sharp arrow point and barb. The arrow was inserted through the cane tube and the leather thong was used to grasp the blunt end of the rod between thumb and finger. The rod was then hauled back, stretching the rubber bands to whatever extent desired. When released, the water supported the arrow, due to its length and speed, and there was very little droop or deflection for a distance of twenty-five or thirty feet. The instrument is only effective in water as in the air the heavy arrow would drop immediately.

Many small sharks hung around our dock, feeding on fish that were thrown back into the water. The dock crew made up some shark tackle and had much sport pulling them out.

One day I saw Mullahey rip open the belly of a freshly caught shark. Its insides spilled on the dock. Mullahey pricked at a large bag-like affair with his knife, and out flopped a baby shark. He tossed the baby overboard. It swam stoutly away.

Mullahey insisted that the small sharks were harmless and

to prove it he often dove overboard and chased them under water with his knife.

One of the dock crew, a chap named Barnes, was very credulous. The rest of the gang had much sport at his expense. They fed Barnes a lot of nonsense about the value of black pearls, and pointed out that they were produced by the type of shell fish that existed on the nearby reef. Barnes fell for this story, hook, line and sinker. Thereafter he spent all his spare time on the reef with a small crow-bar prying open the big mussels and searching for black pearls. Realizing that he was in earnest, the gang had Mullahey shoot a fish with black eyes. They pried one out and planted it in a mussel where Barnes would be sure to find it. He could hardly believe his good fortune. The gang congratulated him and told him that it was worth at least a hundred thousand dollars. Barnes hurried with his treasure to camp, wrapped it in paper and hid it in his bunk. The gang knew that that fish eye would shrivel in a day or so, and to prolong the joke someone stole it. Poor Barnes was inconsolable.

Junius Wood interviewed Barnes about this affair and made copy of it for one of his press stories. Junius cultivated Barnes and the latter expressed an ambition to write poetry. Junius encouraged him to try his hand, whereupon Barnes spent the next three days composing the following masterpiece:

SKYWAY TO ASIA

Commander Grooch in your calm courteous likeable way
Through storm, sunshine, night and day
Commanded the expedition which has set in motion
Swift and safe air travel across the Pacific Ocean.
It has paved the way for greater friendship between East
and West.
For this accomplishment you deserve the best.

This remarkable bit of literary handiwork was posted on the board and was the subject of many irreverent remarks from our hard-boiled crew who swore that Barnes was endeavoring to work himself into a soft job.

Junius Wood spent most of his time ashore and watched our progress closely. His attire consisted of a large straw hat, short drawers and shoes, and he carried his water bottle slung over one shoulder and his camera over the other. His bald head, scanty attire and mahogany shade of tan moved the crew to nickname him "Mahatma Gandhi."

Wake Island was a fisherman's paradise. The lagoon was packed with many species, most of them unknown to us. The reefs swarmed with eels, some as thick as a man's leg. They appeared to live on small fish and spider crabs, and had an uncanny ability to hide under rocks where there seemed to be no opening whatever.

The hermit crabs were queer fellows. We never grew tired of studying their habits. These crabs live ashore and are born without a shell. As soon as they are able to toddle they locate a shell discarded by some other type of shell fish and fit most of themselves into it. At the approach of danger they immediately clew up into a ball inside this shell. From time to time as they increase in size they shift to a larger shell, but at no other time do they leave their borrowed home. It was a novel sight to see a parade of shells moving over the ground with only the red legs of the crabs protruding. Several times I picked up a shell and by violent shaking managed to dislodge the crab who appeared stunned by his misfortune and remained inert until I set his shell down, whereupon he promptly scuttled back into it. The hermit does not have pincers like other crabs and his only defense is his shell. He sticks to it as if glued there.

We were curious to know how each crab was able to find a shell of the right size when growth made a change necessary. The following explanation has been vouched for by eye-witnesses. A dozen or so crabs, ranging in size from baby to almost full grown, line up in a column. In front of the leader is an empty shell slightly larger than his own. At a signal from the leader each crab vacates his own shell and dashes into the one ahead. Thus one empty shell of the proper size makes it possible for the whole dozen to change.

The chief enemy of the hermit crab is a small ground bird known as the wingless rail, which originally hailed from Australia. The rail is a meek-looking little chap, dusty brown in color, a bit larger than a sparrow, and a doughty fighter. I saw one attack a crab with an approach so rapid that the crab did not have time to clew up in his shell. The rail led to the crab's chin like lightning several times and it was all over for the crab. Whereupon the rail and his friends proceeded to eat the hermit.

While testing the well on Peale we pumped five or six hundred gallons of water out on the ground where it formed a large shallow puddle. Dozens of rails took advantage of this opportunity to have a bath. They splashed about and appeared to enjoy themselves hugely.

There were no gooneys at Wake Island and no love birds, but there were large numbers of black terns which made their nests in the trees. They resented any approach, and would swoop down at us, scolding vigorously.

The bo'sn birds were also present. It was their nesting season. They made their nests on the ground and argued and squabbled among themselves in an unmannerly fashion. They often became so interested in their argument that they paid no attention when we approached, so we would take advantage of their absorption and swipe some long red tail feathers for souvenirs. It was easy to locate a nest by the noise coming from the vicinity, to steal up unnoticed and yank out a couple of tail feathers. Whereupon the thief would receive the most blistering abuse from the outraged parents. Perhaps the bo'sn bird gets his name from his ability to imitate a hard-boiled bo'sn swearing at his crew.

During the second week at Wake Island we received a message that made us all feel pretty low. Bill Young had died in the Naval hospital at Guam.

CHAPTER SIXTEEN

PEALE ISLAND began to look quite important as the buildings went up. The station was laid out in the form of a crescent, looking down on the lagoon. The dazzling white sand on the beaches extended out until the water reached a depth of six feet or more. From there on the bottom was coral. The lagoon was so well protected that the water was always quiet. The varicolored bottom gave rise to a many-hued effect in the water, and it only needed a few graceful palm trees nodding in the breeze to complete a perfect tropical picture.

The soil on Peale Island proved fertile. We planted trees, gardens, vines and flowers. There was very little rain at this season but the well furnished an ample supply of brackish water for irrigation.

It was planned on a subsequent expedition to build comfortable hotels for air passengers at Midway, Wake and Guam. Wake Island, due to its many unique features, may become a popular resort. It is far from worldly cares, with a healthy climate and wonderful fishing. The tired business man who wants real rest and relaxation can count on a complete absence of any form of temptation at Wake. Russel, who helped to promote Miami in its boom days, waxed lyric over the possibilities.

SKYWAY TO ASIA

The greatest handicap at Wake was the lack of a proper channel into the lagoon. To construct such a channel would be a long and expensive undertaking as it involved blasting a passage through the solid coral reef. We gave much thought to the idea of a channel, surveyed several likely locations on the reef and estimated the approximate cost. But the job was too big for us.

In the center of the lagoon the water was about fifteen feet deep. There was a sufficiently large area for our planes to land and take off. But the presence of occasional coral heads made it necessary to do considerable blasting and clearing before it would be safe for the *Clipper* to visit Wake Island.

The water at Peale beach was shallow for quite a distance out. In order to service the *Clipper* we needed a dock four hundred feet long. We had not planned on so long a dock and did not have sufficient lumber for it. The nearest port where we could obtain it was Manila. This worried me, for although the work was proceeding at top speed it was still doubtful if we would be able to finish the job within the time limit of the *North Haven's* charter.

The commissary department had its own troubles at Wake. Too late we discovered that our number one cook was a pastry cook. He could make grand pies and cakes but we couldn't live on pastry. None of the other Chinese could cook. So we called for volunteers from the crew. For once they displayed a commendable modesty.

They would try everything else under the sun, but cooking was out. After much earnest solicitation one man admitted that he could cook. He pointed out, however, that he only knew family cooking. Feeding sixty hungry men was a different racket altogether. All hands agreed that if he would undertake the job there would be no complaints. Any man who criticised the food would go on kitchen police immediately.

Our new cook was a bit shaky at first but he improved rapidly. It was evident that his wife had brought him up properly. He indoctrinated the Chinese cook and pulled us out of a bad hole. The crew would take hardship and danger in their stride, but they'd go Bolshevik quickly on poor food.

The tractor was badly needed on Peale Island. In order to release it as soon as possible I decided to unload the gas and oil cargo by hand. We laid some rails across the rocky beach on Wilkes, rolled the drums from the dock over these rails to a cache, and soon sent the tractor over to Peale.

Captain Borklund's temper was getting a bit frayed by this time as he greatly disliked the rather precarious situation of his ship at the reef. Passing squalls sometimes caused the ship's stern to swing sharply in toward the shore. When this occurred there was not time to heave up the anchor. The ship was in danger of grounding and only by going ahead quickly on the engines and dragging the anchor bodily off the reef was she able to get clear. When the anchor slipped off the reef it plunged straight down into the deep blue water, taking up the slack in the chain with an awful thump. The anchor weighed five tons and such a jerk might part the anchor chain and lose the anchor, not only an inconvenience but also a great seagoing crime. Under these conditions it was not surprising that the captain was anxious to get away from Wake Island. I slept aboard ship each night in order to be sure no time was lost in getting an early start each morning. I spent some time each evening with the captain, smoothing him down and listening to his yarns.

He had been for some years a pilot on the Alaskan coast. He was a blunt seaman, not given to romancing, and had no gift of imagination. When he told of some close shave he told it in a seaman's language. He was a vivid story-teller. Unconsciously he took me to a far north country where hardship and danger, treacherous currents, ice-hidden rocks and boiling seas are the customary hazards that face sea-faring men. He told of piloting ships through long, narrow passes on the Alaskan coast, where the current raced so fast the ship could barely make head-way against it. The fog was often so thick that the pilot on the bridge of the ship could not see the bow. On either side the flinty cliffs gave back the echo of the ship's whistle which was the pilot's only guide. Staring ahead through the fog the pilot kept his whistle blowing and the friendly echoes warned him to steer off if the ship veered from the safe channel. According to the captain this sort of navigation was reasonably accurate. If the echo from the cliff on his right reached him before the echo on the left, he shifted the rudder a touch until both echoes reached him simultaneously, which showed that he was in the center of the channel.

It struck me that hanging on to a reef at Wake Island was child's play compared to such an experience. But the fact remained that while he could regard an iceberg with calm confidence, that coral reef gave the captain cold chills.

After thirteen days of high-pressure work we had all cargo ashore except the provisions and frozen meats. We waited four days more before the cold-storage plant was in commission on Peale Island, but on the morning of May 29 they reported ready. The ship made ready to sail.

We left the entire construction crew and the permanent base crew at Wake Island. The base crew for Guam of course went with the ship, as did Taylor, who was to handle the construction at Guam; Junius Wood, Russel, John Steele and I.

McKenzie, Borger and Bicknell were left in charge at Wake. They swore they would have the job finished when we returned from Manila and Guam.

On May 29, at 7 P.M. the *North Haven* gave three long blasts of her whistle and headed west. The lights of the little colony on Peale Island faded slowly from view. Soon they also would signal "ready" to Alameda. Before many weeks had passed a great winged ship of the trans-Pacific air line would settle down on the lagoon, and lonely Wake Island would be lonely no more.

CHAPTER SEVENTEEN

THE *North Haven* was quite light now that the greater portion of her cargo had been discharged, and rode high out of the water. Tail winds and following seas drove her along toward Guam. On the voyage we sighted no ships, not even a fishing boat. Trade routes from the States to the Orient follow a great circle course to Japan, then to China, then down to Manila, and the return is made over the same route. Guam is far off the beaten track. Only one regular freighter, the *Stanley Dollar*, calls there.

No one aboard the *North Haven* had ever been to Guam but we knew that it had some twenty thousand inhabitants, was governed by the United States Navy, had formerly belonged to Spain and became the property of Uncle Sam during the Spanish-American War. Communication in those days had been non-existent at Guam except for itinerant trading schooners. When a United States warship appeared in Guam harbor and fired several shots as a signal for surrender, the Spanish Governor, knowing nothing of a war, sent his aide out to apologize for not returning the salute, stating that the fort was out of powder. So Guam was taken with little effort.

As we neared Guam we passed close aboard Rota Island,

one of the Marianas Group under Japanese Mandate. Through our glasses we could see that the island was inhabited and green fields were in evidence.

My instructions directed me to report to the Naval Governor of Guam. As it is a closed port, which vessels cannot enter or leave, nor passengers land at without permission, I radioed for authority to enter the harbor, and requested a pilot. As we were due to arrive after dark the Governor directed me to lie off until morning because the narrow entrance to the harbor made a night passage dangerous. At daylight a Naval officer, Lieutenant Blanch, came out in a launch and piloted us into Apra Harbor where we tied up to a large buoy.

Apra Harbor is formed by a coral reef on one side and land on the other. The reef serves as a natural breakwater and is adequate for average weather, but it was obvious that big seas would wash over it and produce rough water in the harbor.

Guam has an area of slightly over two hundred square miles. The country is very hilly and quite tropical with coconut palms and dense undergrowth along the beaches.

Shortly after we tied up, the port officials and the Governor's aide, Commander Linsley, came aboard. Commander Linsley stated that before any of our party could go ashore I must pay my respects to the Governor, furnish him with a list of our men, and request permission for them to land. After a short conference, we stepped aboard his launch and were taken through a narrow dredged channel for several miles to the land-locked Navy basin which has excellent protection against typhoons, and is therefore the boat depot for all small craft.

From there we went to the Governor's palace in the town of Agana, where I was formally presented to the Naval Governor of Guam, Captain George A. Alexander, U. S. Navy. He received me cordially, and treated me as if I were still in a Naval status. I delivered to him a letter from Mr. Trippe, the president of Pan American Airways, which served as my credentials, and outlined the arrangements he had made with Naval authorities in Washington for our planes to stop at Guam. The Governor was already familiar with the general plan, and was keenly interested in the project. He assured me that we could count on his staff for every possible assistance, and immediately granted permission for our party to land. This permission was radioed to the *North Haven,* while I made arrangements for tugs, lighters and stevedores to unload our Guam cargo. On the Governor's staff I discovered several former shipmates who placed their cars at my disposal and made me feel very much at home.

When the immediate business of Agana was completed, I drove to Sumay, where our station was to be located. The United States Marines had formerly operated a squadron of seaplanes at Sumay. The hangar, shops and storerooms were still there and in a fair state of repair. Since the Government had no immediate need for these buildings they had granted us permission to use them as a temporary air base.

Sumay was located on the landward shore of Apra Harbor opposite our ship. I found several of our party at Sumay when I arrived.

Just behind Sumay several companies of Marines were quartered. Toward the harbor entrance the cable company had a relay station quite similar to the one at Midway.

We had permission to make minor alterations in the buildings at Sumay and to install our own radio. This work was done with native labor as we had brought only an engineer and a boss carpenter with us. There were no messing or housing facilities at Sumay, so we made arrangements with the cable company to mess our ground crew and to rent us eight rooms in one of their unused dormitories.

The commanding officer of the Marines was a fine chap and very hospitable. He had comfortable quarters and during our stay at Guam often insisted that we come up for lunch or dinner.

The first day Commander Linsley invited me to lunch at the Officers' Club in Agana. The arrival of strangers at Guam is an unusual event. Practically every officer on the Governor's staff was present. They plied me with questions about the transpacific air line.

After luncheon I visited the Naval hospital where the senior surgeon told me of the circumstances of Bill Young's death, and of their efforts to save him. The body had been embalmed, awaiting our arrival. I arranged to have it placed in our chill room for further transfer at Manila to a liner for the States.

With Marine and Navy assistance our Guam cargo came ashore in jig time.

I watched a tow coming in the slip and was surprised to see that the tow boat was a small Navy steamer of a type that was in use on battleships and cruisers previous to the advent of the gasoline motor launch. There were several of these steamers at Guam. They must have been over thirty years old but were still plugging along, emitting clouds of black smoke.

The native Guam stevedores were little fellows, constantly jabbering like monkeys in their native dialect. They were not very strong and it was necessary to hire at least twice as many as was customary in other ports.

Guam is only a short distance north of the equator and the weather is very warm ; so after a full day ashore I was glad to get back aboard ship for a shower and fresh clothes. The Governor had asked me for dinner and sent his boat over early, so I had to hurry.

At dinner at the Governor's house that night I again met his staff and their wives. The Governor made a little speech welcoming our company to Guam, and in conclusion asked me to say something about our trip and our plans.

Extemporaneous dinner speeches are not easy for me but

these people were friendly and genuinely interested in our show, which would bring excitement, frequent mail and variety into their lives. For once I found it easy to talk and I did so for half an hour on the exploits of Pan American Airways in general and of the great North Haven Expedition in particular.

The men at the table were all sea-faring, and readily understood our problems. The ladies were so tired of listening to their husbands' yarns that they were eager to hear anything new. However, it was interesting to note in that secluded port, that the girls were, as usual, more deeply interested in the subject of styles and new dresses. Those who had just arrived on the *Henderson* from the States were looked over carefully and closely questioned as to fashions.

After dinner we adjourned to the Officers' Club to dance. There we were joined by some of our party from the *North Haven* and some of the cable-company staff. It was a good evening and gave me a touch of homesickness for my old Navy days.

CHAPTER EIGHTEEN

Upon inspection I found that the buildings at Sumay, with slight repairs and alterations, would serve our purpose well. There was fresh water and telephone service, but little electric power. However, the Navy told us that they had a new generator on order and would soon be able to furnish us with all the power we could use.

The slip leading to the seaplane ramp in front of the Sumay hangar was very narrow for our big planes. The slip was a hundred and eighty feet wide and nine hundred feet long with stone walls on either side. As our planes were one hundred and thirty feet in span it was clear that they could only be led into the narrow slip safely in quiet weather. But this slip was the only protected location in the harbor and we had to be able to use it regardless of water conditions. So we made plans to install a guide-rail trolley on each wall. The plane would taxi or be towed into the slip, then carefully centered and secured by lines to the trolleys. Then the whole assembly could be safely moved through the slip.

While the cargo was being unloaded, our chief mechanic for Guam fitted up his shop and storeroom. Russel and the radio gang ran all over the island seeking a suitable location for the radio direction-finder. I found myself with several days to spare, and proceeded to find out what I could about Guam.

There is no business on the island except the exporting of copra, the meat of the coconut, used in soap-making. The Naval government has put in schools, roads and sanitary arrangements, and takes a great interest in the welfare of the natives. Each native family owns and operates a small farm in order to be self-supporting. Even those with jobs at the Navy yard or in Agana must work their farms and their duties are arranged to provide the necessary time.

Certain of the more responsible natives sit in council with the Governor from time to time to advise him as to local conditions.

Governor Alexander is a short, stocky, red-headed chap, brimful of energy, so sincerely interested in improving conditions for the natives that he requested a year's extension of his tour of duty there in order to follow up plans he had under way. He had encouraged the natives to adopt English but with indifferent success. They prefer their native tongue and have no need for English except when conversing with Naval officers.

The most difficult problem the Governor had to figure out was what to do with a native boy after he had received an education. The only native business consists of a few retail shops in Agana, and a market, where the clerks are women. After a boy fattens his pride by learning algebra and geometry there is small incentive for him to study.

However, the natives are a happy-go-lucky lot and their future is not a matter of great concern to them or anyone else, except the Governor.

There are practically no horses on the island; the common beast of burden is the water buffalo. It is also quite customary to see a cow ambling along the roads hitched to an antediluvian two-wheel cart. The natives are fond of rice, usually imported because most of them are too indolent to grow it. Some do grow rice, and the soil and climate are ideally adapted to rice culture. At the time of our visit the Governor was planning an irrigation project that would remove any excuse the natives might have for further importations.

A native marriage in Guam is an elaborate affair and the marriage feast or "fandango" lasts for three days. Huge quantities of food are prepared and the guests arrive in large numbers. The ritual requires the bride and groom to sit down at each guest table and taste each platter of food. Such a marathon is enough to give even a Guam native indigestion for a month. One of our party attended a fandango. He stated he had a very interesting evening and enjoyed most of the food but was puzzled to identify a bit of dark rich meat, evidently some sort of fowl. He was told the native name but it was not until the following day that he learned the fowl in question was a large bat which is regarded as a delicacy.

Guam appears to be a healthy climate and there is an entire absence of the native diseases and body sores so common to many of the Pacific Islands.

The Governor's restless energy demanded an outlet and on week-ends he usually took a twenty-five-mile hike over the hills to see what the boys in the back country were doing. His efforts were entirely commendable, but so long as Guam lacks transportation it will remain a backward country. The passage of a transpacific air line through Guam may serve to wake it from the centuries of lethargy that appear to be its heritage from the days of the Spanish conquerors. But the natives will not be impressed.

History has it that when the Spaniards conquered Guam they killed off most of the native men and kept only the women. There is little in the appearance of the present-day native or anything else to mark the passage of the Spaniard except the Catholic religion, which is universal.

The Saturday night hop at the Officers' Club was a gala affair. The *Stanley Dollar* had arrived in port that morning, bringing a few official passengers and a new cable-company superintendent to relieve the incumbent. Her skipper told me he had stopped at Wake Island and discharged a small amount of cargo for our people there. It was an occasion.— When the music stopped a band of the unquenchables sat around and swapped yarns, sang Navy songs and made merry until 2 A.M.

The Governor's two charming daughters, one a blonde,

the other a red-head, liked the party but were bored with Guam because there were no bachelors there, and besides, they wanted to live in Hollywood. Dan Vucetich entertained us with Russian and Turkish songs. Everyone agreed that Dan was a gentleman and a scholar, and would be at home in the Court of St. James's.

Several of our ship's crew had such a grand time ashore that they failed to return to the ship for several days. Captain Borklund asked the Chief of Police to round up the delinquents and hold them for us until just before sailing time. The Chief of Police was a hard-boiled Marine lieutenant. He welcomed this slight diversion. He had no politicians to answer to. When the deserters yelled for a lawyer, he told them that shysters were barred at Guam and if there was any more back-talk he'd put them on a bread-and-water diet. They came aboard, well chastened, before our departure.

Before sailing, our airport manager at Guam was directed to secure for us a hundred tons of fertile soil and have it sacked and ready to load on our return. This soil was for use on Midway in starting a vegetable garden. We also left orders for sprouting coconuts, vines, plants and papaya seed to be planted at Wake and Midway.

We departed from Guam for Manila on June 10 at 1 P.M. Halfway to Manila we received a radio message from Guam stating that the *Pan American Clipper* had again flown from Alameda to Honolulu and then had taken off for Midway. A later message in answer to our anxious inquiry, told us that, with Sullivan in command, she had landed safely in Midway lagoon.

As the *North Haven* pitched over the swells Dan asked me if I thought the gooney birds had given Sully and his crew an appropriate welcome.

CHAPTER NINETEEN

OUR main reason for visiting Manila was to purchase additional material for Wake and Midway. We were well
behind schedule and had to hurry if we were to get back
to San Francisco within the ship's charter limit. So we
decided to stay in Manila only three days. Steele, Russel
and I divided the list of materials to be purchased into
three sections ; each of us agreed to purchase his share.

The ship was now practically empty. Luckily we had
fine weather as she would have had a bad time in a storm.
The typhoons which harry the coasts of the Philippine
Islands, China and Japan, are born in the vicinity of Yap,
an island which lies several hundred miles to the south
and west of Guam. We crossed the typhoon sector with
a fair breeze behind us and a following sea. We reached
the San Bernardino Straits, and threaded our way through
numerous unlighted islands to Manila.

Our ship's brokers there were the Cia General de To-
bacos. Their agent came aboard to discuss our plans. We
told him of our need for haste and he at once agreed to
see that there was no delay in getting our cargo loaded, or
in clearing the ship. We had to load some two thousand
drums of gasoline, and port regulations required this to
be loaded out in the stream away from the docks.

It was unbearably hot on the ship after we docked, so we moved our headquarters to the Manila Hotel.

The firm that was to act as our purchasing agent in Manila was Davies and Company. Their representative, Mr. Gordon, called on us soon after our arrival. We showed him the list of goods we desired to purchase. Mr. Gordon is English. He glanced through the list.

"How soon must you have these supplies?" he asked.

"We sail in three days," I answered.

He shook his head. "You must remember you are in the Orient. Some of these supplies must be manufactured. You will be fortunate if you sail in ten days."

"We can't afford that much time," I said. "The *North Haven* is already behind schedule. We face a big penalty if we fail to return her to California within the charter limit."

His reaction was not enthusiastic so I tried a different tack. I pressed the button and ordered whiskey and soda.

"Mr. Gordon," I said, "we must have this material if the transpacific air line is to start on schedule. The first clipper cannot come to Manila until we finish the bases. There isn't a merchant in town who won't speed up his work if you tell him he is delaying the arrival of the first transpacific air mail. We sail in three days, win, lose or draw."

The Englishman sipped his drink. "It might be done," he said.

He got busy and called up half a dozen contractors and merchants who came to see us.

Russel had to have plans drawn for a radio building at Guam before he could order the material. He got an architect on the job and made him promise to work day and night. I needed three large steel mooring buoys for Guam, and persuaded a contractor to finish them in time.

The following morning we proceeded to Davies and Company's office with our lists. Each of us was assigned a man to show us around and assist in the selection of materials. For three days we did some fast and furious shopping.

At one jobber's I wanted to buy a large quantity of rope. Manila hemp is the finest in the world, but all the jobbers in Manila and the rope factories did not have enough stock of the proper sizes to fill the order. The factories turned to and proceeded to manufacture what we needed.

Even the Japanese and Chinese merchants in Manila seemed to know all about us and our need for haste. They were so anxious to help that at times they did not even stop to bargain, thus foregoing much pleasure and violating the invariable custom of the Orient.

At noon of the second day a serious obstacle arose. The port authorities, in going over the *North Haven's* papers, had discovered that she was overdue to be fumigated. Shipping laws require that a vessel be fumigated every six months. The authorities ruled that we must move the *North Haven* out to a mooring in the harbor, and carry out the regulations. This fumigation business would mean at least two days' delay. I located our attorney, James Ross, who took the matter in hand and arranged that the *North Haven* could delay fumigating until she returned to California.

The morning papers had told us that the *Pan American Clipper* had left Midway the previous day, and made a smooth flight to Honolulu. She had taken off for Alameda after an overnight stop. Later dispatches informed us of her safe arrival. We were beginning to get a routine feeling about long over-water jumps.

Several of us found time that evening to visit "Santa Anna," the largest dance hall in the world, located in the outskirts of Manila. The management furnishes partners, cute dusky maids, with a weakness for flashy clothes and an abhorrence for shoes. The management insists upon shoes. Between dances the girls slip them off and rest their weary feet on chairs.

Manila is a colorful place, with its old walled city in one corner of the modern town, canals, bridges and funny pony hacks on the streets. The lovely old Manila Hotel rests on the harbor front. Its veranda dining room and dance floor are the rendezvous for most of the parties. The Army and Navy Club, several blocks down the waterfront, is famous around the world for hospitality and good cheer.

High up in the mountains to the north of Manila, away

from the heat of the city, is Baguio, favorite summer resort for folks from Shanghai and Hong Kong. The mountains around Baguio are full of gold but the ore is low grade. For years the mines in this region did a mediocre business. Suddenly Uncle Sam hiked the price of gold and those mines went into high gear. Baguio became a bonanza town over night.

Manila was full of political unrest, due to the impending changes of government, which would come with Philippine independence. The average native had the idea that the foreigners would have to move out and leave their houses, lands and chattels behind, whereupon the natives would divide the spoils. An Army officer's wife told me that her houseboy had blandly stated he expected to take over her house when she left.

Early the third morning the *North Haven* moved out to a mooring and began to load drums of gasoline. As our newly purchased goods arrived at the dock they were placed on board a lighter and sent out to the ship. By eight o'clock that night everything was on board.

Steele, Russel and Cushman were to remain in Manila to put in a radio station and make other necessary preparations to receive the first clipper. They saw us off and reminded us to mail them a first flight cover.

We slipped mooring lines and steamed out of Manila Harbor at 9:30 P.M., June 19, and headed back on the long sea trail to the States.

CHAPTER TWENTY

THE first leg of our return trip was enlivened by the presence of Mr. Womack, Civil Service foreman at Guam, who had come along with us from Guam on a vacation trip. At the last moment he had decided to return on the *North Haven*.

Womack is a Spanish War veteran. He told many tall stories about the early days of our occupation of the Philippines. As we passed Corregidor Island at the entrance to Manila Bay he stated that he had helped to build its fortifications, and told me how the place was laid out.

Womack was around seventy years old but hale and hearty. He knew the Philippines and Guam like a book, speaking most of the native dialects. I liked him and enjoyed his stories. A Naval officer at Guam who had known him a long time told me a curious tale about Womack.

Early in the twentieth century Womack had been in charge of a salvage job off the Philippine coast. Several native divers had been down without results and Womack decided to put on a diving suit and go down himself. He became entangled in the wreckage and signaled for help. A native diver went down and succeeded in releasing him but was himself entangled and drowned. Womack looked up the diver's family, and took over their support. He has maintained them ever since.

The morning after we left Manila our ship's crew were a sad-looking lot. The first mate's face was swollen and one eye was nearly closed. His story was that he had left a movie and taken a taxi back to the ship. While *en route* the taxi had been halted by thugs and he had been beaten and robbed. That was the mate's story and he stuck to it. He was a grand seaman but a bad story-teller.

Passing eastward through San Bernardino Straits we came out into open water and found the wind and sea dead against us. We made slow time.

Two days before arriving at Guam we had trouble with the cold-storage plant which had been installed before we left San Francisco. This plant was separate from the ship's cold storage. It was packed full of frozen meats and perishables for Wake Island. A leak had developed and we lost a lot of refrigerating gas. We finally got the leak repaired, and pumped in the last of our reserve gas.

I radioed our airport manager at Guam to secure permission for us to land and to have tugs and lighters standing by when we arrived, so that we could start unloading cargo without delay. We steamed into Apra Harbor, tied up at the buoy at 8 A.M. and went to work.

The airport manager was delighted to receive a new station-wagon we had bought for him in Manila, and a good-sized electric ice box. I called on the Governor and had a pleasant visit.

SKYWAY TO ASIA

At Sumay I inspected the radio station and found the work proceeding satisfactorily; the refueling facilities were installed at the head of the slip and shop and store room were ready to function. The airport manager had taken numerous soundings in the slip and found that in many places it was too shallow to receive the clippers, which draw around four feet of water. The Navy had a dredge and blasting equipment and agreed to dredge the slip for us at cost.

In a conference with the airport manager and the construction engineer, I went over the plans for further improvements and gave them written instructions as to the laying of anchors and the handling of the clippers in and out of the slip.

There were stacks of radios from Alameda to read and answer. I had a busy day. By nightfall we had finished unloading the cargo for Guam which included about twelve hundred drums of gasoline.

However, we still had to load the hundred tons of soil which had been placed on the dock in large sacks. This soil had been rained on several times and was thoroughly wet, which increased the weight of each sack to about three hundred pounds. The little Guam stevedores had a bad time of it. It took three of them to roll a sack onto the lighter. They wanted to knock off for supper. I knew if they did they wouldn't come back until the next day so I had Scott feed them on the ship. They were

[145]

wonderful trenchermen and nearly cleaned the *North Haven* out of food. After their huge supper they grew sleepy and kept sneaking off for a nap. The boss stevedore made the rounds with a flashlight, hauled them out of corners and chased them back to work.

Scott had radioed Guam for coconuts, watermelons and alligator pears for the boys on Wake and Midway. The Governor, anxious to increase trade, had told the market men to be prepared to supply us with as much fruit as we could use. We took half a shipload.

The next morning, June 27, we steamed out of the harbor at daybreak and headed for Wake Island.

The second day out our cold-storage plant began to give trouble again. The refrigerator expert, Rankin, and I tore down the compressors and found leaks in both. We repaired the leaks, pumped in some spare gas we had picked up at Guam and got the plant going again. But the third day it again broke down. Salt water got into the line. That finished us. Scott was able to cram the meat, butter and eggs into the ship's cold-storage plant, but there was no room for the fruit. We spread it in a cool place where it would get the most ventilation, but most of it spoiled.

I radioed Wake Island that we were behind time and could only afford to stay there two days and to make all possible preparations to expedite the discharge of our cargo. We bucked head-winds and seas all the way from

Guam to Wake and made slow time. We arrived off Wake before daylight, July 3, and the wind being favorable, the captain eased up and dropped his anchor on the reef at 5:30 A.M.

CHAPTER TWENTY-ONE

THE first lighter came alongside shortly after we anchored at Wake, bringing the personal effects of the men there who were to return to San Francisco with us. McKenzie came on board and we had a short conference as to the procedure for the next two days.

The ship's cold-storage plant was over-burdened. Some of the meat had thawed enough to indicate that further delay in getting it into the freezing room ashore would be dangerous. The meat and provisions were therefore sent over first.

The big launch, permanently assigned to Wake, had been left at a buoy at the boat landing as there was no way of getting it into the lagoon. It was in good shape, however, as there had been continuous good weather and it had been serviced regularly.

The Wake crew were burned almost black by the sun, and were glad to see us. The dock landing stage and railroad on Wilkes Island were exactly as we had left them, but the camp, with hands and all material had moved from Wilkes to Peale, so Wilkes was a deserted village.

The light launch took us over to Peale and landed us at the long seaplane dock which was complete except for flooring. The dock was built of redwood poles; the legs were stuck into empty oil drums filled with concrete. The seaplane refueling equipment was ready to function. The power house and radio station were complete. The only uncompleted work consisted of such jobs as screens, gutters, painting and plumbing on various buildings.

McKenzie, working against time, had driven his crew to the limit to get the work done. This was quite an accomplishment as the men were on a monthly salary, were a pickup crew for this job only, and had no incentive to work hard. It had been very hot in the sun and there was practically no shade. Some of the men had lost twenty pounds; others had boils and body sores. But they had all done their jobs — thanks to McKenzie's leadership.

The hermit crabs on Peale had proved valuable as scavengers. The rats, however, were quite troublesome. Before we arrived at Wake we had heard many wild stories about the huge red rats which infest the island. One yarn had it that they were so hard pressed for food that they fished with their tails. It was related that a rat would sneak up over a rock and allow his long tail to droop down into the water. An unwary crab would fasten on to the tail with his pincers and the rat would then jerk him out onto the rock and eat him.

In reality, the rats on Wake are grey and about the size of the average American rat. In the beginning we saw very little of them but gradually the word was passed around ratdom that there were good pickings to be had at our camp. Soon the soap began to disappear from the wash-rack. Then one night a half-finished box of candy was devoured. Before long all the rats on the islands had migrated to Peale to investigate the real-estate boom.

The crew took advantage of the rats' passion for candy to have some fun at the expense of Ward and Stuhrman. These two gentlemen bunked in the same tent and each appeared to have difficulty in adjusting himself to camp conditions. Consequently they were the target for jokes. One night while movies were in progress one of the crew sneaked into their tent and tied a small muslin bag of candy under each of their bunks. The bags were high enough off the floor so that the rats would have to jump for them. About half an hour after "lights out" Ward was awakened by a swarm of rats running across his bunk. He let out a yell of terror and reached for a flashlight. The floor of the tent was covered with rats trying to get at the candy. They climbed over each other, jumped on the bunks and indulged in all sorts of antics in their frenzied efforts to capture the elusive swinging bags. They paid no attention to the flashlight.

Stuhrman, awakened by all this, saw Ward standing on his bunk with a flashlight in one hand and a walking stick in the other, striking wildly at the rats and screaming, "Get away, get away, damn you, go away!" He joined the battle at once, yelling lustily for help. The crew had been lying in wait outside for some time. They broke into roars of laughter. McKenzie was awakened by the racket and the yells for help, and went down to investigate.

He finally managed to remove the candy, shoo the rats out and restore order to the camp.

The solar hot-water heaters, which were part of the station equipment, had proved a great success. When they were first hooked up the Chinese boys were amazed to discover that they had ample hot water with no apparent source of supply. They searched the house and grounds for a boiler, never suspecting the black glass-covered case which was the solar heater.

The solar heater, as the name implies, collects the heat from the sun's rays, and applies it to a useful purpose. Our solar heaters consist of a shallow case about twelve feet long, four feet wide and six inches deep. The bottom of the case is lined with heavy black paper which collects the heat. A dozen lines of one-half inch water pipe are fastened over the black paper, four inches apart, and connected together at the ends. This set of pipes is connected to a two-hundred-gallon tank which is mounted adjacent to the solar heater. Circulation is caused by the fact that warm water rises. The case of the solar heater is covered with heavy window glass, placed so as to receive the direct rays of the sun during the larger part of the day. The heat of the sun's rays is held and diffused by the heavy black paper, and the glass cover keeps this heat imprisoned.

The crew quarters at Wake consist of two six-room bungalows with showers and lavatories. Each room is furnished with a bed, dresser, table and chairs. The airport manager has a separate bungalow with two bedrooms, living room, kitchen and bath. There are separate quarters for the Chinese. There are also a large shop, storeroom and power house, two windmills and a pump house. All quarters have large screened porches, hot and cold running water and electric light.

The operations office and radio receiving station are in the same building. The radio transmitter is in a separate building, and is operated by remote control from the receiving station. The radio direction-finder office is set up in the center of the radio antenna poles.

As I headed back for the ship late that evening, after thoroughly inspecting Wake, I reflected that perhaps never before had a lonely desert island been transformed so quickly into a modern little city.

CHAPTER TWENTY-TWO

Due to several delays, the work of unloading at Wake had progressed slowly. I realized that to finish up the next day we would have to work faster than we yet had.

That night after most of the men from Wake returned aboard ship, I had the steward serve each man a big drink of whiskey and a cold bottle of beer with supper. Loud cheers came from the crews' quarters.

I called the gang bosses up to my cabin and told them that we had to sail the next night as an extra day at Wake would cost the company a heavy penalty. They promised to put the rest of the cargo ashore before darkness on the following day. They did!

During the day I went back to Peale Island for a final conference with the airport manager and the chief mechanic on how to handle the clippers when they arrived. We walked out to the end of the long seaplane dock where the seaplane float would be moored. Here the clippers would be refueled and serviced.

A three-inch pipeline ran from the big reservoir tank at the head of the dock to the outer end. This pipeline would be connected to the gas pit on the float by a flexible hose. Gasoline for the clippers would be pumped through the pipeline to the gas pit. From the gas pit an additional hose would lead to the clipper tanks. This arrangement would permit the clippers to refuel at the rate of one hundred gallons per minute.

We went over the exact procedure to be followed in handling the clippers, and made notes. Then we inspected the landing and take-off area in the lagoon. The regular launch was busy with cargo so we boarded a queer-looking little craft named the *Kewalo*. She had been sent out from Honolulu on the *Stanley Dollar*. The *Kewalo* was a complete washout. She was underpowered, the rudder was so small that she had poor maneuverability, and she was so poorly balanced that extreme care was necessary at all times to avoid capsizing.

The *Kewalo* was Borger's pet hate. One day Borger was out on the *Kewalo* with several of the crew when the engine quit. While the crew were working on the engine the tide and wind drifted the boat toward the reef. One of the crew, noting the danger, yelled at Borger to throw the anchor over. The *Kewalo* had an anchor but no one so far had bothered to arrange for an anchor line, and the bare anchor was lying on the bow. Borger, being no part of a seaman, failed to notice that the anchor had no line and simply picked it up and threw it overboard. Instantly he realized his mistake but it was too late. The heartless boat crew laughed at his attempts to explain. Two of them jumped overboard and fended the launch off the rocks until the engine was started. When they returned to Peale Borger's ears burned for a week.

Quite a bit of blasting had been done in the lagoon and many coral heads removed. But the space cleared was still too restricted to give a safe take-off distance for the clipper ships. While the *Kewalo* circled the lagoon I pointed out to the airport manager where the limits of the take-off area should be marked and directed that the entire area inside those limits be blasted clear of coral heads, and then the entire area dragged to check the presence of any obstructions.

We returned to Peale about noon. I watched the airport manager take his upper-air soundings for his routine weather report to Alameda.

Upper-air soundings are made to determine the direction and velocity of the winds at varying altitudes. They would be especially valuable to the captain of a clipper in flight, because it is entirely possible that there may be a head wind on the surface and a favorable wind at ten thousand feet, or *vice versa*. The soundings are taken from a balloon about two feet in diameter, painted red and filled with hydrogen. The balloon is released and as it ascends, its bearing and vertical angle are measured at regular intervals with a theodolite. The theodolite is fitted with a telescope which can follow the balloon to high altitudes. The operator of the theodolite simply follows the balloon with the telescope and a clock mechanism on the instrument automatically records its altitude, direction, and horizontal travel. The data thus obtained is radioed in at regular intervals from all stations to the weather forecaster at Alameda and is invaluable to him in making up his weather forecast for the entire transpacific route.

To finish the construction work at Wake we decided to leave Borger, George Kuhn and six of the construction crew behind. They would later be returned to San Francisco by one of the clippers. Mullahey liked the place and wanted to stay, for which we were thankful as he was our key man for the blasting of coral heads in the lagoon.

Mullahey was a fine diver and could stay under water for nearly two minutes. The blasting operations intrigued him. He would dive down and inspect the base of each coral head for a crevice in which to place the dynamite. Then he would come to the surface, take down several sticks of dynamite, and pack them into the crevice. Next he would take down the ignition wires and connect them. Then he would climb into the boat which would pull away a couple of hundred feet and "let her go." Shooting fish and shooting dynamite were both fun to Mullahey. Incidentally, the blasting operations killed quantities of fish. Many new species were noted.

After a final glance around the station, all hands returned with us to the ship to say goodbye. The unloading crew had finished their work. The ship was ready to sail.

It was agreed that as soon as we left, all water-front gear would be snugged down, because good weather could not last forever. The big launch had to be left moored in the open sea. She was to be fitted with a storm bridle and serviced as often as possible until we could make better arrangements. The landing stage and the spare lighter were to be ferried across the reef and through Wilkes Channel into the lagoon. The permanent ground crew who were to remain at Wake consisted of the airport manager, chief mechanic, power-house mechanic, launch mechanic, two radio operators and six Chinese.

There were many last-minute exchanges of messages for friends and the folks at home. At 5 P.M. we heaved the anchor off the reef and headed for Midway. The big launch raced alongside of us for several miles, blowing her Klaxon and imploring us to expedite the arrival of the clipper with mail. I gazed at the tiny coral atoll as it receded in the distance, and wondered if the clipper would safely find her way to that lonely little lagoon so far from friendly ship lanes.

CHAPTER TWENTY-THREE

AFTER leaving Wake Island someone remembered that it was July 4. In honor of the day and departure, a cup of grog was issued to all hands. The crew was happy to be homeward bound.

We had three hundred and fifty drums of gasoline, one thousand sacks of cement, one hundred tons of soil and incidentals for Midway. But there was only one launch and one lighter there and it was a long tow from the northwest passage to the beach. I radioed the airport manager to try to borrow the cable company's lighter to help speed up our unloading, and to have all waterfront facilities prepared for our arrival. Unless we caught bad weather at Midway we were now reasonably sure to arrive back at San Francisco within the ship's charter limit.

We arrived off Midway early on the morning of July 8. The airport manager came out in the launch through the southeast passage with the cable company's lighter in tow. He reported that conditions at the reef were unfavorable and advised against attempting to tow through the passage. We put a small load on the lighter and directed the ship to move over to the northwest passage. I went in on the launch with the lighter in tow, through the southeast passage. We were lucky to get through as the breakers were beginning to be dangerous.

We dropped our tow at the beach, picked up our own lighter and headed for the ship. The Wake crew showed little enthusiasm for discharging Midway cargo. No one had helped them do their job at Wake — why should they help Midway? But the Midway gang had work on the station to finish so the Wake outfit had to handle the unloading. They never got out of second gear.

The cable company's lighter was twenty years old and could not stand much banging about. We had it nearly loaded when the continued battering of the swells caused the seams to open and it began to sink. We loaded the material back onto the ship with all possible haste, meanwhile pumping the lighter out. But water leaked in faster than we could pump. By the time we had the lighter unloaded it was awash. We towed it ashore and beached it. A couple of our carpenters went to work caulking it and finally made it water-tight. We didn't use it again.

The air base at Midway was virtually completed. The seaplane dock was three hundred feet long as it was necessary to run out that far to get five feet of water for the clippers. At the shore end of the dock was a substantial workshop built on piles to protect it from high water. The various buildings and facilities at Midway were similar to those at Wake. The Midway crew were in good shape.

As previously stated the *Pan American Clipper* had arrived at Midway in our absence. The airport manager stated she had been handled and serviced without difficulty. Naturally there had been much excitement at Midway. The crew there were not air men and flying the Pacific still sounded like a Jules Verne tale to them. Every half hour *en route* the Clipper had given her latitude and longitude to Alameda, and the radio operator at Midway caught it also. The airport manager marked these positions on the big wall chart. While still hundreds of miles away the Clipper had made direct radio contact with Midway.

She had arrived at 2 P.M. Dan Vucetich had been right. The gooneys and terns had been almost as excited as the Midway crew. The clipper circled the lagoon. Captain Sullivan saw the Midway launch flying the checkered flag in the blue-water area, squared away and landed alongside. The launch led him up to the landing float where the Clipper was secured.

The crew came ashore for their first close-up of Midway and received a hearty welcome from all hands. They found Midway interesting, were fed, saw a good movie, and turned in. In three hours the base crew serviced the plane and secured it for the night. No air man could ask for more.

She shoved off the following day on her return trip to Alameda.

Subsequent to the Clipper's arrival, a fleet of long-distance Naval planes had flown to Midway from Honolulu on a tactical problem, and had anchored for ten days in the lagoon. Several warships were anchored outside. The island was a busy place.

I called on Mr. Perry, the cable company superintendent, and found him with a broken leg which he had suffered in a soft-ball baseball game staged by the cable company versus our crew.

He told me another yarn. While the Navy was visiting Midway, there was an exchange of hospitality between the ships and the island, and several admirals came ashore. At that time the terns were nesting and laying quantities of eggs. At dinner, one of Perry's guests, an admiral, expressed a curiosity about tern eggs, and said he would like to try one. Later in the evening as the guests were leaving, Mr. Perry presented the admiral with a bottle of Scotch, done up in a square package. Thinking the admiral might be embarrassed in the presence of his officers if the contents of the package were revealed, as Navy ships are dry, Perry told the admiral, with a wink, that it contained tern eggs. Apparently the admiral missed the wink. The next day his aide told Perry when the admiral arrived at his ship he gave the package to his mess-boy and told him to fry the contents for breakfast. The next morning the mess-boy had great difficulty convincing the admiral that no one could possibly fry the contents of that package, though he might get fried on it.

The young gooneys on Midway were now almost fully

grown. Great numbers of them were attempting to learn to fly, and the beach was loaded with "student pilots." They made exactly the same mistakes that the student aviator makes, only the results were not serious. When a young gooney attempted to take off cross-wind, he would start to skid as soon as he had flying speed, and usually wound up against the side of the nearest sand dune. Or he would take off successfully into the wind, circle the beach and come in to land down-wind, and promptly go up on his nose and over onto his back. After each such accident the young gooney would struggle to his feet and shake the sand off himself with a most disgusted attitude. This show was going on all over the beach. I never saw anything funnier.

The parent gooneys had departed on their annual migration which was a cruel thing for the youngsters, as none of them had as yet learned to feed themselves. The parent bird feeds its young by regurgitating its food in semi-liquid form, holding it in its wide-spread beak while the youngster inserts his beak and sucks up the food. This method of feeding teaches the youngster nothing about how to fish for himself. When the parent bird leaves, he must learn to fly and to fish entirely by instinct.

When he first learns to fly his feathers are still downy and get water-soaked easily. If he lands on the water his feathers get so wet that he can't take off. As a result thousands of young gooneys drift around the lagoon and are gobbled up by sharks or drift out to sea and are lost.

Our crew spent their spare time at Midway gunning for man-o'-war birds. These fellows are cut-throat sky bandits who prey on other birds and steal their food. A man-o'-war is a sea hawk with talons. Only part of a man-o'-war's food is got by robbery. The rest he gets by hunting for himself. When he sees a tern or a booby scoop up a fish, he promptly takes it away from him.

The cleared space in front of the mess hall was jammed with thousands of nesting terns, crowded shoulder to shoulder on the sand; each bird facing the wind. So far as we could discover, no bird on the island laid more than one egg per season. The Midway fowl are poor folk and work hard for a living. They don't go in for large families.

CHAPTER TWENTY-FOUR

IT took us four and a half days to get our Midway cargo ashore but at last we were ready to sail. Dan Vucetich finished checking his inventories and reported that the only piece of goods unaccounted for on the entire trip was one pot of red lead.

The permanent ground crew to be left at Midway was identical in composition with that of Wake. We also left three of the construction crew behind to finish the odd jobs. We said goodbye to the gang at the cable company at noon on July 12. The Midway crew took us out to the *North Haven* in the launch and saw us off.

We had ample time to stop at Honolulu long enough for the ship to take on fuel oil, drop the Honolulu crew, and still make San Francisco within the charter limit. As the gang on the *North Haven* had been away from civilization for several months, I knew that they would turn Honolulu inside out if they had all night there. So I decided to play it safe and arrange our arrival too late for them to go ashore. The weather and head winds assisted this scheme beautifully. With only a small change of pace, we dropped anchor off Honolulu at 10:00 P.M.

The crew suspected me, and checked the engine-room log to see if we had slowed down. But I assured them, as all could see, that only the heavy northeast trade had held us back.

Shortly after we anchored, our airport manager for Honolulu, Van Zandt, came under our stern in a launch and asked us to heave him a line so he could pass up some mail. Almost everyone got a letter and I got a thousand dollars in cash. I had radioed ahead for this so that I could give each man a few dollars with which to purchase trinkets.

The port officials came out early, and we docked at Pier 26 at 8 A.M. I suggested to the captain that as we were only stopping long enough to refuel, it would be safer not to advance his crew any money, which was sure to go for hard liquor. He agreed with me but was afraid that if he refused to pay off, his crew would report him to their union when they got back to California as being a hard skipper.

Before the ship docked I advanced a few dollars to each man in the construction crew and told them we were going to sail promptly at 2 P.M. Anyone who had too good a time would probably miss the ship. I would be very, very sorry to leave them in Honolulu but the ship would sail at 2 P.M.

McKenzie and I went ashore to inspect the temporary base for the clippers. This is in Pearl City, adjacent to the Naval Air Station. It is a lovely spot.

The property consists of a lot bordering the waterfront, with a large bungalow set well back from the waterfront.

With its wide verandas and huge living room, it would serve as an operations and traffic office until we could make permanent arrangements. Waterfront conditions were ideal for handling the clippers.

After leaving there we called on several local officials to discuss a municipal seaplane base. We realized that it would be unwise to select a permanent location in Honolulu until we had had sufficient operating experience there with the big planes.

We returned to the ship shortly before 2 P.M. A delegation from the Naval Air Station was waiting to see us. They were part of the crowd who had been at Midway and were all former shipmates of mine. We swapped ideas about Midway and the future of military and commercial aviation.

We were interrupted by a rumpus on the dock and in the forward part of the ship. I stuck my head out a porthole. Our construction crew had made a valiant effort to drink all the beer in Honolulu, and being afraid they would miss the ship had come back early and brought their friends, and their beer, with them. They were singing the songs of all nations, dashing up and down the dock, on and off the ship, yelling like Indians. Each man had a couple of leis around his neck. The general effect was festive to say the least. The dock policeman was greatly amused at their antics. I had no objection as it was perfectly natural for them to celebrate their return to civilization.

The Naval crowd went ashore. I took a turn around the ship. As I had feared, practically all the ship's crew were too drunk for duty. The only sober men were those on engine-room watch, and the third mate. The first mate was barely able to walk.

The pantry cook had decided he liked Honolulu and was going to stay there. He was a smallish man with a wooden leg. He lurched up and down the dock mumbling threats at the captain because he had refused to pay him off. McKenzie and I, by dint of much persuasion, had managed to get all of our crew aboard, and were anxious to hoist in the gangplank, but we had to have that cook.

I asked one of our husky foremen, named McCullough, who happened to be sober, to run down, pick the cook up bodily and bring him aboard. McCullough sauntered up to the cook, grabbed him, and started for the gangway. But the cook knew a trick or two. He reached down and applied a punishing hold. McCullough let out a yell and dropped him. A dock bystander ran up to lend a hand. He took the cook under one arm while McCullough took the other, and they again started for the gangway. The cook took hold of the stranger's hair and wrenched with all his might, whereupon the stranger took a poke at the cook's jaw. Unfortunately the blow landed high and laid open the cook's cheek, as the stranger was wearing a heavy seal ring. Wagner, a Honolulu boy who had been with us at Wake, saw this and decided that no stranger had a right to slug our cook. He promptly ran over and knocked the stranger cold. We finally got the cook on board and into the first-aid station.

The third mate ran the winch while our crew hauled in the gangplank, manned the lines and got the ship away from the dock. Then the *North Haven* staggered out of the harbor.

The ship's crew had cached a supply of liquor on board. We were two days at sea before all hands sobered up.

After leaving Honolulu we bucked into the northeast trades and a rough head sea. While *en route* home Mc-Kenzie and I got our reports in order, while Dan Vucetich audited his accounts so we could pay off all hands immediately upon arrival at San Francisco.

We made slow time with the ship high out of the water and arrived at San Francisco early on Sunday morning, July 28, just four months after we had steamed through the Golden Gate outbound.

The port authorities quickly inspected us. We docked at Pier 22 shortly before 8 A.M.

A large crowd was waiting for us on the pier. The police had erected a barrier to prevent them from rushing on board.

Our division manager, Colonel Clarence Young, came on board at once. Behind him were the newspapermen and their photographers. They interviewed us for half an hour while the cameramen shot us from all angles. The press stories spoke of us as "unsung heroes."

Our paymaster from Alameda came aboard and was soon

engaged in paying off the men on the after part of the ship. Each man had a fat check when he said goodbye. The dock was a hubbub as they joined their friends and families. Mother, wives and sweethearts laughed and wept as they clung to their men back from the dangers of roaring seas, hidden reefs and shark-infested waters.

The Ancient and Royal Order of Gooneys were invited to a spread at the Alameda Airport the following night. There Colonel Young expressed the company's appreciation of their efforts. He told them that the men of other nations might build transoceanic air lines, but their efforts had made it possible for America always to say, "We did it first." The gang leaders made speeches. All hands were proud to have had a share in blazing the Transpacific Trail for the *China Clipper*.

CHAPTER TWENTY-FIVE

THE *North Haven*'s affairs completed, I reported to Alameda Airport armed with a collection of photographs illustrative of the various phases of our work. All department heads gathered for a week-long conference. McKenzie and I proceeded to enlighten them on what had been done and remained to be done at our Pacific bases. Honolulu and Midway had already received a clipper plane. Wake and Guam would be ready shortly.

Alameda Airport had not been idle during our absence. It had become a combined laboratory and training school. Pan American had assigned some of their best men to the key positions in the Transpacific division. They realized that flying the Pacific was the biggest job any aviation company had ever undertaken. No expense or pains had been spared to secure expert organization and direction.

I was assigned to the operations department. The operations manager, "Dutch" Schildhauer, is an Annapolis graduate and an ex Naval aviator. Dutch had helped organize the Pan American pilots' school in Miami and was well qualified to supervise training activities in the Pacific. He was pushing the training program at top speed in order to have flight crews indoctrinated when the transpacific air-mail schedules started.

The function of the operations office is to dispatch all planes and direct their movements. To do this successfully it must keep in constant contact by radio with all planes and stations, and provide them with accurate weather and other information.

Schildhauer's office was equipped with several interesting gadgets. An enlarged chart of the North Pacific covered most of one wall. The chart had a glass cover upon which notations were made with red chalk. Each day this glass cover was marked to show the positions of steamers along the course, and the weather they had reported by radio. Above the chart stood a row of electric clocks, one for each station in the division. Each clock was labeled and kept the local time of the station it represented. Above these was a large clock keeping Greenwich time. To avoid confusion all division messages would be sent on Greenwich time.

Weather is a major consideration with any air line. In the transpacific division it is exceptionally important because of the very long non-stop flights involved. A special weather bureau is necessary to secure accurate weather forecasts. At Alameda this department was next door to the operations office. In charge was Bill Clover, formerly chief weather forecaster for Transcontinental and Western Air.

Bill discussed with me some of the obstacles his department faced. He had been used to forecasting weather in the States where there was a weather-reporting station every twenty-five miles. In the Pacific he must depend on weather reports from our widely separated airbases and from occasional ships along the route. Beyond Honolulu there were practically no ships along our air route as it deviates from the steamer lanes. And only a few of the Pacific steamers were equipped with radio sets of sufficient range to keep in touch with San Francisco.

Pan American Airways had set in motion plans to secure additional reports from ships at sea. They arranged to have certain ships equipped to take upper-air soundings. Special reports were arranged for from certain U. S. and foreign weather bureaus. When these plans matured Bill Clover could not only forecast storms or bad weather along the airway but also plot a course for the clippers that would provide favorable winds and the best possible flying conditions.

Across the hall in the chief pilot's office I found Captain Sullivan in charge. He and I have been friends since Navy days when we flew in the same squadron.

"Sully, what do you hear from Musick?" I asked.

"He's testing the *China Clipper* at Baltimore and says she's a honey."

"When will she be ready for the job out here?"

"Not before early in November."

"I hear on your next flight you're going through to Wake Island."

"Sure. As soon as those bozos clear a few more coral heads out of the lagoon."

"What do you think of our radio direction-finders?" I asked. "Do they click?"

"They're swell," he answered. "We gave them a real test on the return flight from Midway to Honolulu and they were a hundred per cent. Right after we left Midway we pulled the hood down in the cockpit and flew blind all the way. The direction-finder at Midway gave us our bearings for the first half of the flight and the Honolulu base held us on the course the rest of the jump. When we pulled up the hood we were right over Honolulu."

"Suppose the radio compass goes haywire?"

Sully grinned. "Oh, the navigator wasn't blind. He knew where we were every minute. If we'd got off the course he'd have stepped in with the drift indicator and his celestial navigation. But he didn't have to tell us a thing."

"That was a good workout," I said. "You'll need a direction-finder when you start looking for Wake Island. It's about as big as your hand."

Sullivan and I went down to have a look at the *Pan American Clipper* (Sikorsky S42). Nearly all the seats had been removed and all but one of the compartments were packed full of gas tanks which gave the plane a range of three thousand miles. It had been designed for the South American trade but made an excellent training ship for the Pacific division.

During the next few weeks at Alameda the *Pan American Clipper* made daily training flights to improve the technique of flight crews. The operations office prepared daily problems in navigation which required the plane to fly several hundred miles to sea on a given course. At the end of this leg of the flight, instructions were radioed to the plane to intercept a ship two hundred miles away. The ship's position, course and speed were given to the plane. The plane crew had to plot the ship's position on the chart, and lay a course to intercept her. After the ship was found the plane was instructed to return on a zig-zag course to Alameda. Fifty miles out the hood was pulled down in the cockpit. This completely obscured the view, simulating a condition of heavy fog. The pilot at the controls had to finish the last leg flying by instruments entirely.

These problems were much more difficult than navigating a straight flight to Honolulu or Midway. Several junior flight officers were assigned to each flight. Their work was supervised by the navigation instructor, Fred Noonan.

In flight, Noonan directed them as to the proper use of navigation instruments. Later he corrected their paper work and pointed out mistakes. Training was tedious work but it was absolutely necessary. All of us realized that while the radio direction-finder was a great aid it was not infallible, and our navigators must be able to find their way without it if necessary.

Captain Sullivan grew weary of the long training grind and said he wished the gang at Wake Island would "shake a leg." He cheered up when Captain Jack Tilton, another ex-Navy man and South American veteran, reported in from Miami and took over a share of the training flights. Shortly after Tilton's arrival Wake Island reported ready.

Sullivan and Tilton made the flight to Wake without the slightest hitch. *En route* they picked up some ice cream and new movie films at Honolulu for the crews on Midway and Wake. When they departed from Midway for Wake we followed every move anxiously as that flight was the toughest test of our navigation methods. Every half hour they radioed their position reports to Alameda via Midway. We plotted these positions on the chart. They invariably showed the plane to be exactly on the course. In the Clipper, Noonan "shot the sun" every hour; his fixes agreed with the direction-finder bearings.

Sully and Tilton hit Wake Island "on the nose." It was a calm day and the water in the lagoon was so clear that every rock on the bottom was visible from the air. Sullivan circled over the lagoon and radioed Wake that there appeared to be only six inches of water. Wake replied there was fifteen feet. Sullivan landed and taxied up to the float.

The Wake crew, in smart white uniforms, welcomed the crew of the Clipper, the first plane to land there. Another section of the air bridge across the Pacific was ready to function.

The Clipper's crew explored Wake, the reefs and the lagoon for two days, and collected quantities of souvenirs.

The base crew serviced the Clipper and inspected her thoroughly. On the return flight to Alameda the Clipper flew at various altitudes to check the speed and direction of the winds aloft.

Training flights were resumed at Alameda almost as soon as she landed. We were preparing for a weekly flight schedule to the Orient which would require six flight crews. Every man must be qualified before the schedules started in November.

The final exploratory flight was to Guam. The U. S. Post Office permitted first flight covers to be carried on this flight and thousands of them poured in from all over the United States.

Sullivan and Tilton flew the *Pan American Clipper* to Guam, via Honolulu, Midway and Wake, without incident. There they were cordially received by the Governor and his staff on October 13. On their return to Alameda they reported that all stations were prepared to handle planes on schedule. The stage was set for the inaugural mail flight of the *China Clipper*.

CHAPTER TWENTY-SIX

At Baltimore the *China Clipper* was rapidly completing her Department of Commerce tests in Musick's skilled hands. These were important because the Post Office Department would not award a mail contract to any company unless the aircraft to be used satisfied their demands for high performance. The tests were satisfactory and Pan American finally had a mail contract for the route they had already pioneered. Shortly thereafter Musick flew the *China Clipper* to Miami. There she underwent further tests including a twenty-five-hundred-mile non-stop flight to San Juan, Puerto Rico, and return. Then Musick flew her to Alameda via Acapulco, Mexico.

She arrived at Alameda on Armistice Day. A holiday crowd jammed the airport and lined the waterfront, watching for her arrival. A dozen planes circled overhead, waiting to greet the newcomer.

The Clipper hove in sight, dropped down to a low altitude and swept across the airport so all could view her graceful lines and gigantic size. Then she circled several times, landed on the bay, taxied into the yacht harbor and up to the float.

The crowd at the airport struggled to get near enough for a good look. They were familiar with the transport planes in common use but this silver flying boat was America's first ocean air transport. Few of them would actually fly to China in her but they wanted to inspect the ship that was going to make such a flight possible.

At the float the ship was cleared of crew and baggage, and the beaching crew took charge. They led her over to the ramp, rolled the beaching cradle down into the water and floated it into position under her hull. The V-shaped cradle is made of steel tubing padded to protect the hull, and is fitted with wheels and flotation gear. A tractor hooked on to the cradle and hauled it, with the glistening Clipper resting on it, up the ramp, and onto the field. Close up, her size was impressive. The wing span is a hundred and thirty feet; from nose to tail she measures ninety-two feet. The crowd milled around her and stared at the great wings high over their heads. Everyone had seen her in the air, yet I heard a dozen people say, "How can anything as large as she is really fly?" I felt a bit that way myself.

However, there were more believers than skeptics. In the next few days the Pan American traffic office in San Francisco received three hundred requests for reservations to the Orient on the *China Clipper's* first flight. Just two years had passed since my English friend in Manila had scoffed at the idea of a transpacific airline. I addressed him an inaugural-flight air-mail cover and dropped it in the mail box.

Alameda Airport was feverishly active as the new ship was tuned up. I looked her over carefully. She is comfortably arranged inside. The lounge has the spacious effect of a drawing room and seats sixteen people. There are three other passenger compartments, each equipped to seat twelve people or sleep six. The seats convert into berths much like those on a Pullman. The sound-proofing is excellent. It is easy to converse during flight in normal tone of voice with a person six feet away. There is ample head room and space for passengers to stretch or walk about.

The *China Clipper* is designed to cruise most efficiently at ten to twelve thousand feet. At that altitude the air is thin and offers less resistance than it does lower down, which permits higher cruising speeds for a given amount of power. The engines are supercharged to operate at high altitudes and the pitch of the propellers can be varied from the cockpit to suit the altitude. At ten thousand feet over the Pacific the clippers will usually be above the clouds and in smooth air, and yet not so high as to cause discomfort to passengers from lack of oxygen. There is an automatic heating device which maintains a comfortable temperature in the cabin.

The four motors are rated at one thousand horsepower each. The plane can take off fully loaded on three of them and fly on two if necessary. The possibility of a forced landing is remote but if it should happen she has the whole ocean to land in. There is no danger of her bumping into a mountain or landing in a swamp.

During flights, the captain of each clipper will be advised by Alameda of the position of every steamer along the route. In an emergency he would notify the nearest steamer, land as close to her as possible and transfer his passengers. The air bases each maintain a powerful sea-going launch equipped with a radio set and direction-finder. These launches will serve as rescue boats if necessary.

The clipper ships, like any ocean liner, are equipped with life-saving apparatus. Instead of heavy wooden life boats they carry rubber life rafts, each made up in a compact package, which contains a flask of compressed gas for inflating the raft, a set of oars, signal flares, emergency rations and fresh water. The gas flask is attached to the raft in such a fashion that it is only necessary to turn a handle to inflate the raft in twenty seconds.

Several test flights were made with the *China Clipper* soon after her arrival to check instruments. Night landings were practiced because bad weather or head winds might force her to arrive at an air base after dark.

Musick and his crew were guests at luncheons and dinners for several days before the inaugural flight. California gave the transpacific air line a big hand from the start. The entire state prepared to celebrate her maiden flight to the Orient.

The departure date was set for November 22. The

SKYWAY TO ASIA

Governor of California proclaimed it Pan American Airways day. A state-wide committee was formed to assist at the birth of America's first transoceanic air line. An international broadcast was arranged to include speakers from Alameda, Honolulu and the Philippines. The world was invited to listen in.

CHAPTER TWENTY-SEVEN

NOVEMBER 22 dawned clear and fair, visibility unlimited, ceiling unlimited. Crowds began to arrive at the airport several hours before departure time. The *China Clipper* was launched and secured at the foot of the ramp, facing the speakers' stand, where the broadcasting apparatus was installed. Four flags were flying. The Stars and Stripes nodded to the territorial flag of Hawaii, the red and blue banner of the Commonwealth of the Philippines, and the personal flag of the Postmaster General of the United States.

A hum spread over the airport as the crowd streamed in and notables began to arrive and mount the speakers' stand. At 2:40 the loudspeakers announced that the international broadcast would start in five minutes. The announcer opened the program:

"Here on the platform are distinguished national figures who are to participate in today's ceremonies. Across the Bay one hundred thousand people are watching and listening on the shores near the Golden Gate. A thousand miles away in Hawaii are other thousands. Eight thousand miles across an ocean, halfway around the world, are still other thousands. On the five continents of the globe are millions who by the magic of radio are about to witness with us one of the most dramatic events in the history of our modern world."

Mr. J. T. Trippe, the president of Pan American Airways, took the announcer's place.

"We are assembled here today to dedicate regular commercial air service across the Pacific Ocean to Hawaii, the Philippines and the Orient, an area extending over one-third the way around the world.

"It is significant and appropriate that the first scheduled air service over a major ocean route is being started under the auspices of the American Government, by an American company operating aircraft designed and built in the United States and in charge of American captains and crews.

"The first long-distance, over-water route flown regularly anywhere in the world was the six-hundred-mile journey across the Caribbean from Kingston, Jamaica, to Barranquilla, Colombia. For five years Pan American has flown this route : it has been our laboratory of preparation for the service instituted today. There our technical staff has developed our ocean direction-finding and navigation apparatus, and there our flight captains and their crews have qualified for over-ocean service. So complete and effective was this accumulated experience than when our first experimental flight across the Pacific was undertaken six months ago, the *Pan American Clipper* cruised without incident over the twenty-four-hundred-mile course and set down in Honolulu one minute late."

SKYWAY TO ASIA

Postmaster General James A. Farley faced the "mike" and read a letter of congratulation from President Roosevelt, then offered his own tribute.

Departure time was drawing near. The crew of the Clipper filed aboard over a narrow catwalk leading to the front hatch. In Navy-blue uniforms and jaunty white caps they were a smart-looking lot. Sullivan led the way aboard. He was assigned as first officer for the flight. Sully had flown the Pacific so often he could recognize some of the waves as old friends. Fred Noonan, the navigator, came next. Fred had been a navigator in the old square-riggers. The crew maintained that he could "shoot the sun" standing on his head. Then George King, second officer, an able young pilot in training. Next were Chan Wright and Vic Wright, not brothers, engineering officers, who knew every bolt and nut in the entire ship and called each engine "pal." Then Bill Jarboe, the radio officer, a wizard with the ether, in charge of communications. Last to step aboard was Captain Ed Musick. Forty-one years of age, he has been flying for twenty-two years. Conservative to a degree, he has never had a serious accident. There is no pilot anywhere as well qualified to command this flight.

The beach crew turned the *China Clipper* around and moved her out to the float. Meanwhile the broadcast continued with congratulatory speeches from the Governor of Hawaii and the President of the Philippines.

When these were finished Mr. Trippe spoke into the

"mike." *"China Clipper,* are you ready?" The loud-speaker answered: "Pan American Airways' *China Clipper,* Captain Musick. Standing by for orders, sir."

"Stand by for station reports,"•ordered Trippe.

The loudspeaker gave voice in rapid succession:

"Pan American Airways Ocean Air Base Number One —Honolulu, Hawaii. Standing by for orders."

"Pan American Airways' Mid-ocean Air Base Number Two — Midway Islands. Standing by for orders."

"Pan American Airways Transpacific Air Base Number Three—Wake Island. Standing by for orders."

"Pan American Airways Mid-ocean Air Base Number Four — Guam. Standing by for orders."

Pan American Airways Transpacific Air Terminal — Manila, Commonwealth of the Philippines. Ready and standing by, sir."

Mr. Trippe reported the readiness of the line to the Postmaster General, who then ordered the inauguration of scheduled service on Foreign Airmail Route Number Four-teen at 3:28 P.M. Pacific Standard time.

The *China Clipper* was instructed to cast off and depart for Manila.

The lines were cast off, engines speeded up and the Clip-per taxied smartly through the yacht harbor and out into the bay. Police boats kept the takeoff area clear while she circled to warm up her engines.

At 3:48 Musick opened her throttles wide. We stood

there watching her, wishing her good luck and happy landings.

Hundreds of cameras clicked from launches lining the course as the *Clipper* raced past them, spurning the water from her sides, leaving a mile-long plume of foam to mark her passage, toward the Golden Gate. She left the water just as she passed under the center span of the Bay Bridge. People cheered ; every ship and ferry in the harbor gave a whistle salute as the great silver ship dipped gracefully in acknowledgment. She flew low across the Marina where the school children of San Francisco were gathered to wish her Godspeed. Then faded slowly into the sunlit western sky, bound for Manila.

CHAPTER TWENTY-EIGHT

AT Alameda Airport the operations office, the weather bureau and the radio department prepared for an all-night vigil as the *China Clipper* speeded westward. The weather bureau received constant teletype reports of the weather over the Pacific and kept the operations office advised of changes. The radio department kept in constant touch with the Clipper and gave the navigator a radio compass bearing every half hour. Ships at sea regularly reported the weather, and their speed, course and location. Musick radioed a position report every half hour which gave his latitude and longitude, altitude, speed and direction of the wind, ground speed, temperature of the air, ceiling, visibility, and local weather conditions. Our training at Miami and Alameda had been so long and thorough that the office routine went off like clockwork.

The weather forecasts for the trip were excellent and checked closely with reports from the plane. Musick climbed to ten thousand feet and set a course slightly southward of the great circle, to avoid heavy cloud-banks ahead. The navigator took sights of the sun every half hour.

Night came on and the stars peeped out. Noonan reported that he felt at ease now, because he could get a more accurate fix from the stars than from the sun.

But suddenly the stars were blotted out by a cloud-bank high overhead, which extended up to over twenty thousand feet. The plane could not afford to waste gas climbing that high so Noonan had to depend on dead reckoning and radio bearings for three hours.

Through a break in the clouds below, Musick spotted the twinkle of a steamer's lights. He sent her a message asking her position. The reply showed the Clipper to be dead on the course but making a bit faster speed than the navigator had figured.

When the high cloud-bank was passed the crew of the Clipper had no further worries. During the night Noonan got seven fixes from the stars and received forty-one radio bearings from Alameda and Honolulu.

At Alameda the newspapers copied the position reports as they came in to us, and these were flashed to the world. The *China Clipper* landed at Honolulu at 10:19 A.M. Honolulu time, after a flight of twenty-one hours, three minutes. We turned in for a few hours' sleep.

There was a day's lay-over in Honolulu to service the plane and rest up the crew. A two-ton cargo of freight for Midway and Wake, which included fresh vegetables and turkeys for Thanksgiving dinners, was loaded aboard the Clipper. The Honolulu postoffice sent down two hundred and seventy-five pounds of mail which brought her mail load for the Orient up to one thousand six hundred and fifty-three pounds. Fourteen passengers were taken on board for Midway and Wake. These were Pan

American personnel and included staff replacements and Chinese cooks and waiters. It was a great day for the China boys. They had never been aboard an airplane before.

That afternoon from Alameda we radioed Musick a weather forecast for the next day's flight and early the next morning we again radioed all weather and other information available. After checking the weather carefully, Musick took off for Midway at 6:35 A.M. The navigator took his departure from a lighthouse on Niihau Island and set the course directly for Midway.

This course follows a chain of islands and reefs which makes navigation easy and affords an opportunity to check the accuracy of radio bearings by land observations. It is the only stretch along the entire Pacific air route that provides any scenery. The rest of the way there is nothing to look at but cloud effects and vast expanse of the sea.

Our weather forecast had advised Musick that halfway to Midway he would encounter a "cold front," which means rain squalls and shifting winds. Seven hundred miles out he saw heavy cloud banks ahead and dropped down under them. His position reports showed that his speed was reduced by contrary winds and that there was poor visibility. But the Clipper barged on through and came out into bright sunshine fifty miles from Midway. She circled over the lagoon and landed at 2 P.M. The base crew made a fast job of servicing the plane and before dark she was moved out to the night mooring ready to fly at daybreak.

We sent her the weather forecast and other dope for the next leg.

The next morning at 6:12 A.M. the *China Clipper* took off for Wake Island. Two hundred miles out she crossed the one hundred and eightieth meridian, the international date line, and Monday became Tuesday. An hour later she passed over a Matson Line steamer bound for Yoko-hama. Radio greetings were exchanged. The Clipper waggled her wings in salute and noted, but did not hear, three white plumes of steam from the ship's whistle.

Halfway to Wake heavy cloud-banks covered the course. The Clipper plunged through them at one hundred and fifty miles per hour, flying blind, by instrument. Musick's reports gave Alameda a clear picture of the flight. In such weather Wake Island would not be visible five miles away. The navigator must depend on dead reckon-ing and radio bearings.

But the Clipper's crew were not worried. They knew that as they neared Wake Island the bearings would change rapidly unless they were on the true course. They had a small direction-finder on the plane with which they took bearings on Wake Island for a double check. Fifty miles from Wake, Musick dropped down under the clouds, five hundred feet above the sea. The bearings held true and Wake Island popped out of the sea, two miles ahead.

Musick's crew were thrilled to see the Stars and Stripes

floating from a tall mast on Wake. The Wake personnel
were equally thrilled to hear that the turkeys had arrived.
A large pasteboard crate proved to contain twenty-five
canaries, a gift from the cable company's staff at Midway.

The *China Clipper* was serviced and tucked away for
the night at Wake. After an early dinner the flight crew
turned in to be ready for a dawn takeoff on the long flight
to Guam.

The next morning there was a dead calm. The *China
Clipper* with a lightened load took off easily at 6:01.
There were favorable winds at low altitude and the Clipper
flew west at twelve hundred feet, under a ceiling of heavy
clouds.

Half-way to Guam she passed over the U.S.S. *Chester*,
eastbound out of Manila. Secretary of War Dern was
aboard the *Chester*. The Clipper radioed him an invita-
tion from Mr. Trippe to inspect our transpacific bases and
to have Thanksgiving dinner with the Wake Island staff.

Noonan wanted to check his position by "shooting the
sun" so Musick climbed through a thick cloud layer and
broke out on top at eight thousand feet. Here the Clipper
found a thirty-mile tail wind that boosted her into Apra
Harbor, Guam, at 3:05 P.M.

Most of the population of Guam were assembled at the
Pan American air base to welcome her. She delivered
twenty thousand letters, a large part of which were first-
flight covers. There was a day's layover at Guam to service the plane and give the flight crew a rest.

At Alameda we worked up the weather map and other data and sent it to Guam by radio.

The following morning at 6:12 A.M. the Clipper took off from Apra Harbor and set a course almost due west for Manila, sixteen hundred miles away. Her position reports were now relayed to Alameda via Guam, Wake, Midway and Honolulu but they reached us within fifteen minutes after being sent from the plane. They showed that the plane was changing altitude occasionally to check the force and direction of the wind.

Musick would drop down to a thousand feet while the navigator dropped a bomb over the side and took a drift sight. The bomb is a thin glass jar containing a quart of aluminum powder. Upon striking the water the glass shatters and the powder instantly spreads, to form a spot twenty feet in diameter on the surface of the sea. By observing this spot with a drift indicator the navigator measures the drift of the plane and estimates the speed and direction of the wind.

The winds were favorable. The Clipper gained an hour and a half in time because she was traveling with the sun. At 2 P.M. the high rugged hills of Luzon Island were a distant shadow on the port bow. Flying high, the *China Clipper* held straight on her course which led across the island, Manila being on the west coast of Luzon. A squadron of Army planes flew out to escort her in. When she landed in Manila Harbor most of the population were assembled on the waterfront. There was so much excitement that Alameda had to wait an hour before the official landing-time was radioed from the Manila base.

The first United States air mail was soon delivered to the Philippine Post Office — forty-five thousand letters. Newspapermen besieged the Clipper's crew and demanded the story of the flight. Musick told them that the flight was "without incident." That was poor material for a newspaper story. The reporters wanted headline stuff: "Clipper lost in fog but battles her way through!" "Clipper almost forced down by storm at sea!"

They were disappointed to hear that fog is no menace to the clipper ships because the crews are trained to fly by instruments all day if necessary; that the clippers may buck heavy weather occasionally to keep schedules, but have no intention of blundering into storms. So the reporters came back to earth and made capital of the fact that the *China Clipper* had flown from California to Manila in fifty-nine hours forty-eight minutes actual air time. They compared this to the fastest steamer time of twenty-one days, and prophesied a renaissance of business for the Philippines.

During the next two days at Manila the Pan American mechanics inspected the *China Clipper* thoroughly and prepared her for the return flight to California. Government officials boarded her and admired her spacious cabins. The Philippine Post Office delivered to her seventeen hundred and eighty-nine pounds of mail for Guam, Honolulu and the States.

The first leg of the return flight was the most difficult because it involved either a night takeoff from Manila or a night landing at Guam. Flying east sixteen hundred miles the Clipper would lose an hour and a half in time. The weather forecast showed a strong head wind. The flight time required would be approximately fourteen hours.

Musick radioed Alameda that he preferred a night takeoff from Manila. Alameda agreed. He took off at 2:53 A.M. and headed due east into a starlit sky. His first position report arrived at Alameda while he was climbing to cross Luzon Island at five thousand feet. But the next one showed that he was dropping down close to the sea to avoid the strong head winds aloft. Flying at five hundred feet most of the way to Guam he bucked an average breeze of twenty-one knots on the port bow.

It was dark when the Clipper landed at 6:41 P.M., Guam time. She had radioed her arrival time and the base crew at Guam had arranged landing lights so there was no difficulty in handling the ship. The flight crew had dinner and went to bed immediately while the base crew worked most of the night servicing the Clipper.

A typhoon was heading north across the airway from four hundred miles to the south of Guam. Its progress was closely watched by the weather bureau at Alameda. We were relieved when the *China Clipper* hauled out of Guam early next morning, just ahead of the storm. The typhoon passed three hundred miles west of Guam. Six hours after the Clipper's departure, Guam was covered by squalls.

The typhoon had stirred up the weather for a thousand miles in every direction. Its low center sucked in the wind from all quarters.

The *China Clipper* bucked through squall after squall of blinding wind-driven rain *en route* to Wake. The flight crew knew that they would run out of bad weather after several hours and simply settled down to see it through.

Hour after hour at Alameda we plotted the Clipper's course from her position reports and worried our heads off at her slow time and the bad weather. We discovered later that she had let the automatic pilot fly the plane through the squally area and that we were the only ones who were excited.

The Clipper landed after dark at Wake Island. There was no difficulty as the base crew had landing lights set and the launch standing by. The flight crew were put to bed without delay as they had to be up at dawn for the flight to Midway.

The weather next day was perfect except for consistent head winds. The junior officers took turns at navigating and piloting, being checked for errors occasionally by Musick and Noonan. The Clipper landed at Midway well before dark and was quickly refueled and serviced.

Midway's time is four hours later than Alameda's so we were able to make up our morning weather map at Alameda, and get a complete forecast through to the Clipper before she took off. Our forecast showed head winds on the surface from Midway to Honolulu. We requested Midway to take an upper-air sounding as soon as the Clipper departed.

Midway sent up a balloon and found that at ten thousand feet the Clipper would have a twenty-mile tail wind. This was radioed to Musick. He climbed up into this favorable breeze and made fast time to Honolulu.

The flight crew had a breathing spell at Honolulu as they laid over until the following afternoon. This gave the base crew a chance to go over the ship thoroughly and prepare her for the long flight to California.

The cargo from Honolulu to the mainland was highly scented. It included large quantities of lovely flowers, some of them addressed through to New York by air mail all the way, and a mail load for the mainland of one hundred and eight thousand letters.

The *China Clipper* cruised at ten thousand feet, toward Alameda, a mile above the clouds, under a clear sky. Our radio operator heard her exchanging greetings with several ships at sea. Noonan got a fix every hour from the stars. Not once did she swerve from the great circle course.

Through the night at Alameda we followed her every move as she sped smoothly along the sky trail. Dawn came and she was only four hours from home. A following wind helped her to break the speed from Honolulu to California. Seventeen hours and four minutes

SKYWAY TO ASIA

after taking off at Honolulu she landed in San Francisco Bay.

Musick taxied into the yacht harbor and up to the float. The crew stepped ashore looking as fresh as they had on departing from Alameda two weeks before. They said the two weeks' experience had been very interesting; they had been cordially received everywhere along the route; that the plane and motors had functioned perfectly; the air bases had given excellent service.

So the inaugural transpacific air-mail flight was completed on schedule.

Before long the route would be extended to China, weekly mail schedules would start, then passengers would be carried.

Larger and faster planes will fly the Pacific tomorrow. Most of them will follow the trail of the *China Clipper*.

APPENDIX
Log of the *China Clipper*

WESTBOUND

	Miles		
L. San Francisco	2410	3:46 P.M. Friday	Nov. 22, 1935
A. Honolulu		10:19 A.M. Saturday	Nov. 23
L. Honolulu	1380	6:35 A.M. Sunday	Nov. 24
A. Midway		2:00 P.M. Sunday	Nov. 24
L. Midway		6:12 A.M. Monday	Nov. 25
	1260	[Time advanced one day crossing International Dateline westbound]	
A. Wake		1:38 P.M. Tuesday	Nov. 26
L. Wake	1560	6:01 A.M. Wednesday	Nov. 27
A. Guam		3:05 P.M. Wednesday	Nov. 27
		[Remained one day according to Original Schedule to Arrive Manila November 29]	
L. Guam	1600	6:12 A.M. Friday	Nov. 29
A. Manila		3:32 P.M. Friday	Nov. 29

Local Times Throughout

Westbound : 8210 Miles — 59 hours, 48 minutes

EASTBOUND

	Miles		
L. Manila	1600	2:53 A.M. Monday	Dec. 2
A. Guam		6:41 P.M. Monday	Dec. 2
L. Guam	1560	6:11 A.M. Tuesday	Dec. 3
A. Wake		8:57 P.M. Tuesday	Dec. 3
L. Wake		6:45 A.M. Wednesday	Dec. 4
	1260	[Time retarded one day crossing International Dateline eastbound]	
A. Midway		4:49 P.M. Tuesday	Dec. 3
L. Midway	1380	6:11 A.M. Wednesday	Dec. 4
A. Honolulu		5:27 P.M. Wednesday	Dec. 4
L. Honolulu	2410	3:02 P.M. Thursday	Dec. 5
A. San Francisco		10:36 A.M. Friday	Dec. 6

Local Times Throughout

Eastbound : 8210 Miles — 63 hours, 24 minutes

Total Flight : 16,420 Miles — 123 hours, 12 minutes

CHINA CLIPPER

CLYDE SUNDERLAND
OAKLAND

NORTH HAVEN LOADING AT SAN FRANCISCO

DISCHARGING CARGO AT MIDWAY ISLANDS

"BREAD LINE" DURING FIRST DAYS AT MIDWAY ISLANDS

TEMPORARY RADIO SHACK AT MIDWAY

BABY GOONEY

CONSTRUCTING MESS HALL AT MIDWAY ISLANDS

LOVE BIRDS, MIDWAY

BABY LOVE BIRD ON NEST

PARENT GOONEYS GOSSIPING

YOUNG GOONEYS ON THE BEACH ON EASTERN ISLAND

CALIBRATING THE RADIO COMPASS. NOTE THE ENGINEER ABOVE
TAKING ACCURATE BEARINGS OF THE *NORTH HAVEN*
WITH A TRANSIT

FIVE-TON GENERATORS GOING ASHORE AT WAKE ISLAND

GOONEY LODGE INITIATION CEREMONIES

PREPARING FOR INITIATIONS INTO THE ANCIENT AND
ROYAL ORDER OF GOONEYS, ON THE *NORTH HAVEN*

CONSTRUCTING A DOCK AT THE BOAT LANDING, WILKES ISLAND

CLEARING A RIGHT OF WAY FOR THE RAILROAD ON WILKES ISLAND

THE RAILROAD ENDING AT THE LAGOON

THE RAILROAD IN ACTION

CLEARING WILKES CHANNEL TO BRING LIGHTER THROUGH

PUSHING THE LIGHTER THROUGH WILKES CHANNEL

TRACTOR IN WILKES CHANNEL HAULING THE LIGHTER
OFF A SHALLOW SPOT

THE LIGHT LAUNCH IS PUT OVERBOARD INTO THE LAGOON

MULLAHEY WITH HIS FISHING SLING AND ARROW

WINGLESS RAIL ATTACKING A HERMIT CRAB

TRACTOR ARRIVING AT PEALE ISLAND

MESS HALL AND COLD STORAGE ON PEALE ISLAND

THE MAIN HIGHWAY—GUAM

FAST TRANSPORTATION—GUAM

LOOKING DOWN ON THE STATION ON PEALE ISLAND FROM THE
TOP OF THE WINDMILL

COMPANY STREET AT PEALE ISLAND

120   160   160   120

RUSSIA

ALASKA

NOME   FAIRBANKS

CANADA

BETHEL

JUNEAU

VANCOUVER

CHINA

PACIFIC

UNITE
STAT.

PEIPING

SAN FRANCISCO

SHANGHAI

MIDWAY IS.

LOS ANGELES

CANTON

PHILIPPINE IS.   WAKE IS.   HAWAIIAN IS.

GUAM IS.

SAIGON

SINGAPORE

NEW GUINEA

OCEAN

SOERABAYA

AUSTRALIA

LEGEND
United States-Pan American Airways
Great Britain - Imperial Airways
France - Air France
Germany - Deutsche Lufthansa
Netherlands - Royal Dutch Airlines
Russia - Soviet Air Trust

120   160   160   120